Isotopes: A Very Short Introduction

VERY SHORT INTRODUCTIONS are for anyone wanting a stimulating and accessible way into a new subject. They are written by experts, and have been translated into more than 45 different languages.

The series began in 1995, and now covers a wide variety of topics in every discipline. The VSI library now contains over 500 volumes—a Very Short Introduction to everything from Psychology and Philosophy of Science to American History and Relativity—and continues to grow in every subject area.

Titles in the series include the following:

Rob Ellam

ISOTOPES

A Very Short Introduction

OXFORD
UNIVERSITY PRESS

OXFORD
UNIVERSITY PRESS

Great Clarendon Street, Oxford, OX2 6DP,
United Kingdom

Oxford University Press is a department of the University of Oxford.
It furthers the University's objective of excellence in research, scholarship,
and education by publishing worldwide. Oxford is a registered trade mark of
Oxford University Press in the UK and in certain other countries

First edition published in 2016
Impression: 2

Published in the United States of America by Oxford University Press
198 Madison Avenue, New York, NY 10016, United States of America

British Library Cataloguing in Publication Data
Data available

Library of Congress Control Number: 2015959274

ISBN 978-0-19-872362-2

Printed in Great Britain by
Ashford Colour Press Ltd, Gosport, Hampshire

Contents

Preface: at home with the Beilbys

The word 'isotope' first emerged in the scientific world in 1913 when Glasgow-based chemist, Frederick Soddy, coined the term in a letter to the editor of the journal *Nature*. Later, Soddy would credit Margaret Todd with suggesting the term derived from the Greek for 'same place' (*isos topos*) to describe the phenomenon whereby substances that differed in mass appeared to occupy the same place in the periodic table of the elements. The story goes that Todd was a dinner guest at the home of George and Emma Beilby, who since 1908 had been Soddy's in-laws. The Beilbys lived at 11, University Gardens (Figure 1), in the West End of Glasgow—a building that survives as part of the University of Glasgow. Todd was a close associate of Sophia Jex-Blake (the Edinburgh 'lady-doctor') and both were leading lights in the woman's suffrage movement. The Beilbys, Soddy, and his wife, Winifred, were strong supporters of the movement, and it is likely that the common connection with Jex-Blake resulted in Soddy and Todd discussing his work and coining a word that still retains its original meaning today.

George Beilby was a prominent industrialist with interests in the chemical industry, including a factory producing cyanide in the Maryhill district of Glasgow. Frederick Soddy appears to have come to Beilby's attention through their mutual association with William Ramsay, with whom Soddy worked at University College,

1. 11, University Gardens as it is today.

London. Beilby, living close to Glasgow University, was well connected with influential people in the university and socialized both with his professorial neighbours and more radical and younger figures on the left wing of politics in Glasgow. Contemporary descriptions suggest that the Beilby household was a hot-bed for both scientific and political discussion and debate. Beilby seems to have been influential in the university decision to create a post of 'Independent Lecturer' to establish a research school in chemistry, and in appointing Soddy to that position. Furthermore, Beilby appears to have funded Soddy's academic set-up 'dowry' in Glasgow by securing the supply of 50 kg of uranium nitrate to kick-start Soddy's radioactivity research. Undoubtedly, Beilby had anticipated the economic potential of radioactivity and would soon be instrumental in the mining of radium at Balloch on the shores of Loch Lomond.

Clearly, Soddy appreciated the importance of the isotope concept in 1913, and he could have been in no doubt when in 1921 he was awarded the Nobel Prize in Chemistry for his work on nuclear transmutation—basically what we now tend to call radioactive decay. There are strong indications that he predicted the military uses to which the power stored inside atoms would be put. However, it is less clear that he could have imagined the myriad constructive uses to which isotopes would eventually be applied.

In the chapters that follow, we will explore Soddy's legacy as applied to Earth, environment, archaeological, planetary, and biomedical sciences. The journey will show how radioactive isotopes are used as the natural atomic clocks that established the enormous magnitude of geological time. We will see how isotopes are used to monitor metabolism in humans and other animals. Our story will show how isotopes can be used to reconstruct the diet and lifestyle of pre-historic people and to complete our understanding of animal dietary and migratory behaviour. We will see how isotopes, even some that no longer exist, have been used to reconstruct the evolution of the Earth and the development of the early Solar System—establishing timescales for the assembly of the pre-solar nebula into the sub-components of meteorites, the meteorite themselves, whole planets, and the differentiation of those planets towards their present-day structures. During the journey we will discover that nature itself constructed the world's first nuclear reactor some 1,700 million years ago. Along the way, we will explore the various measurement technologies that generate the actual isotopic measurements that facilitate these exciting and fundamental applications. We will discover instruments of exquisite sensitivity capable of finding atoms so rare that their detection and quantification is comparable to identifying an individual leaf in a whole forest. Welcome to the world of isotopes.

Acknowledgements

I am grateful to Gordon Cook and Rob Mairs who read parts of the manuscript covering subjects in which their expertise greatly exceeds my own. I am indebted to those who influenced, encouraged, and shared a life spent exploring isotopes and Earth science. You are too numerous to mention individually, but Peter Hopwood started my geological journey in high school; my parents, Derek and Margaret Ellam, supported my university aspirations; and Tony Fallick gave me an early opportunity and much encouragement. My wife, Rowena, and children, Jenny and Cammy, provided love and support throughout. Finally, I dedicate *Isotopes* to my grandson, Oliver McAlpine, whose gestation corresponded with, but was rather shorter than, the time it took me to write the book.

List of illustrations

List of illustrations

Chapter 1
Identical outsides . . . different insides

> Put colloquially, their atoms have identical outsides but different insides.
>
> (Frederick Soddy, Nobel Prize acceptance speech, 1922)

With this colloquialism, Soddy was describing what we now tend to know as the Rutherford–Bohr model of the atom: a nucleus surrounded by a cloud of electrons. Today the model has been replaced by the discovery of a host of other sub-atomic particles (most recently the Higgs boson) but to understand isotopes the Rutherford–Bohr model is close enough. Let's take the simplest atom, hydrogen, which incidentally constitutes about 75 per cent of the mass of the universe. Most hydrogen has a nucleus containing a single positively charged particle or proton and, in orbit around the proton, a single negatively charged electron. We call these 'charged particles' but the charges are tiny, 1.6×10^{-19} (that's 0.00000000000000000016) coulombs. To put this in context, the smallest fuses commonly in use in domestic electrical systems in the UK are set to 'blow' when the current exceeds 3 amperes (amps). An amp is a coulomb per second, so to blow the 3 amp fuse takes about twenty million trillion (2×10^{19}) electrons per second—there are a lot of electrons and a lot of atoms in the universe. Protons and electrons are also very small, a proton weighs about 10^{-27} kilogrammes and an electron is about 2,000 times lighter than a proton.

Hydrogen and carbon isotopes

Most hydrogen is made of one proton and one electron, and occasionally gets called protium. However, about one atom in every 6,420 contains an extra particle, a neutron. The neutron is the same size and mass as a proton and it sits in the nucleus of the atom with the proton. However, as its name suggests, it carries no charge and its effect on the element is negligible for everything except mass. The nuclear configuration of the elements is represented in standard chemical notation by atomic numbers and atomic masses. The atomic number, expressed as a subscript numeral, is the number of protons. The atomic mass, expressed as a superscript numeral, is the total number of protons and neutrons. So, normal hydrogen is ^1_1H but hydrogen with a neutron added (deuterium) is ^2_1H (i.e. one proton, one neutron, and one electron). Deuterium is just like hydrogen in all its physical properties, for example an odourless, explosive gas, but twice as heavy; which is, of course, still much lighter than air. Deuterium oxide or heavy water looks and smells just like normal water, although those extra neutrons lead to some important applications in nuclear science. So, in Soddy's colloquialism, protium and deuterium have the same 'outsides' (a single electron) but different 'insides' (nuclei with either, (H) a proton or (D) a proton plus a neutron); they are both 'isotopes' of hydrogen. Furthermore, both atomic configurations are stable, so they are 'stable isotopes' of hydrogen.

There is a third isotope of hydrogen, tritium, which has one proton and one electron (those are what make it hydrogen) but tritium's nucleus has two neutrons. This is an unstable nucleus which undergoes radioactive decay (to which we will return shortly) to the element helium. So, we call tritium a radioactive isotope and the helium produced by its decay is a radiogenic isotope. The average lifetime of a tritium atom is about twelve years, so any tritium incorporated into the Earth when it formed is long gone. However, interaction between the Earth's atmosphere and cosmic rays produces small amounts of tritium and it is an occasional product of spontaneous nuclear fission. Mostly though, the small amounts of tritium in the atmosphere are by-products of the explosion of nuclear weapons and emissions from nuclear power generation.

Let's now look at another important element, carbon. Most carbon has six protons, six electrons, and six neutrons—$^{12}_{6}C$ and is known as carbon-12. However, about 1 per cent of carbon has seven neutrons—$^{13}_{6}C$—carbon-13. ^{12}C and ^{13}C are both stable isotopes of carbon. About one in a trillion (million million) carbon atoms has eight neutrons in its nucleus (i.e. $^{14}_{6}C$). This configuration of six protons and eight neutrons is unstable, so carbon-14 is radioactive and decays into nitrogen. The lifetime of ^{14}C is quite long though and ^{14}C or 'radiocarbon' is the basis for the radiocarbon dating technique that will be introduced in Chapter 2.

Radioactivity and radioactive decay

Having introduced the idea that some isotopes are radioactive, it is timely to explore the different types of radioactivity. We recognize four styles of radioactive decay. α, β, γ (alpha, beta, and gamma, the first three letters of the Greek alphabet) and spontaneous fission. Overall, radioactivity or radioactive decay is the process by which the nucleus of an unstable atom loses energy by emission of ionizing radiation. An isotope that emits radiation spontaneously is called radioactive. At the level of a single atom, radioactive decay is a random (or stochastic) process and it is impossible to predict when an individual atom will decay. However, the chance that an individual atom will decay is known and remains constant through time. For a large number of atoms, we can determine the decay constant: the rate at which radioactive decay will proceed.

Another way of looking at this is in terms of 'half-life', which is the length of time needed for half the radioactive isotopes of a substance to decay. Let's imagine a million radioactive atoms with a half-life of one year. After a year there will be half a million radioactive parent isotopes left and half a million new daughter isotopes generated. After a further year, there will be quarter of a million parent isotopes and three-quarters of a million daughters. Another year, and it will be 125,000 parents and 875,000 daughters.

The numbers involved are arbitrary, the point is that over one half-life the parents will halve and the daughters will double, and this is true for any snap-shot of the decay history (i.e. after one year, the reduction of the parent is half that present at time-zero but similarly after eighteen months the number of parents would be half those present six months after time-zero). Systems that decay or grow like this are called exponential, but for radioactivity the decay constants and half-lives differ greatly. Some isotopes are so unstable that they last for only fractions of a second and the evidence that they ever existed comes from the presence of their daughter products or by watching them decay in real-time—in a nuclear reactor for instance. Other isotopes are much longer lived with half-lives that are longer than the age of the Earth. ^{238}U has a half-life quite close to the age of the Earth (4,500 million years), so half of the ^{238}U that ever existed on Earth has already decayed into other elements, ultimately isotopes of lead. Samarium-147 ($^{147}_{62}Sm$), which is widely used as a chronometer and tracer by geologists, has a half-life of about a hundred billion years, so only about 4.5 per cent of ^{147}Sm has yet decayed into its daughter neodymium-143 ($^{143}_{60}Nd$).

α-decay

Let's now look in more detail at the different types of radioactive decay remembering that all will proceed as exponential processes.

α-decay is also called 'heavy particle decay'. In fact, the 'heavy α-particles' are not so heavy, being merely the nuclei of helium atoms composed of two protons and two neutrons. Nonetheless, α-decay is a mechanism by which to lose protons and neutrons quickly, allowing heavy isotopes to decay to more stable lighter configurations. The $^{147}_{62}$Sm to $^{143}_{60}$Nd decay mentioned earlier is an α-decay. Notice that the atomic number (subscript) reduces from 62 to 60 reflecting the loss of two protons while the atomic mass (superscript) has reduced by four to represent the total of protons and neutrons lost. Sm and Nd are so-called rare earth elements or lanthanide metals and you can see that the decay of ^{147}Sm to ^{143}Nd shifts two spaces to the left in the periodic table (Figure 2) reflecting the loss of those two protons.

A couple of interesting asides here: first, notice that there is an element, promethium (Pm), located between Nd and Sm, but in our (and most other) periodic tables Pm is represented differently to the other lanthanides. This is because Pm has no stable isotope and all its radioactive isotopes have short half-lives. Therefore, no natural Pm has existed since about a hundred years after the formation of the Solar System. Of the other elements, leaving aside the transuranic actinides (those to the right of uranium), only technetium (Tc) is naturally extinct. There is Pm and Tc on Earth but only through their creation in nuclear reactors and weapons. Second, for years lanthanide rare earth elements like Sm and Nd lingered in the 'nether regions' of the periodic table with little other than academic interest. Recently though, compounds of neodymium, iron, and boron were discovered to have extremely strong magnetic properties. Suddenly, the rare earths are big business where tiny but strong magnets are needed—indeed, you are likely to be carrying NdFeB magnets in your mobile phone to make the speaker work or drive the vibrate function.

β-decay

β-decay is also known as 'isobaric decay'. In this context, isobars are isotopes with the same atomic mass, so the β-decay of $^{14}_{6}$C into $^{14}_{7}$N

1 H																	2 He
3 Li	4 Be											5 B	6 C	7 N	8 O	9 F	10 Ne
11 Na	12 Mg											13 Al	14 Si	15 P	16 S	17 Cl	18 Ar
19 K	20 Ca	21 Sc	22 Ti	23 V	24 Cr	25 Mn	26 Fe	27 Co	28 Ni	29 Cu	30 Zn	31 Ga	32 Ge	33 As	34 Se	35 Br	36 Kr
37 Rb	38 Sr	39 Y	40 Zr	41 Nb	42 Mo	43 Tc	44 Ru	45 Rh	46 Pd	47 Ag	48 Cd	49 In	50 Sn	51 Sb	52 Te	53 I	54 Xe
55 Cs	56 Ba	57 La	72 Hf	73 Ta	74 W	75 Re	76 Os	77 Ir	78 Pt	79 Au	80 Hg	81 Tl	82 Pb	83 Bi	84 Po	85 At	86 Rn
87 Fr	88 Ra	89 Ac	104 Rf	105 Db	106 Sg	107 Bh	108 Hs	109 Mt	110 Uun								

Lanthanides	58 Ce	59 Pr	60 Nd	61 Pm	62 Sm	63 Eu	64 Gd	65 Tb	66 Dy	67 Ho	68 Er	69 Tm	70 Yb	71 Lu
Actinides	90 Th	91 Pa	92 U	93 Np	94 Pu	95 Am	96 Cm	97 Bk	98 Cf	99 Es	100 Fm	101 Md	102 No	103 Lr

2. Periodic table of the elements—elements in grey text (e.g. Pm) have no naturally occurring isotopes.

is isobaric. In β-decay a neutron-rich nuclide (nuclide is intended to imply the nuclear configuration of an isotope) spontaneously transforms a neutron into a proton and an electron. The electron is then emitted as a β-particle. The net effect of β-decay is slightly counter-intuitive because the 'decay' actually produces a new proton so the daughter has a higher atomic number (A) than the parent isotope (i.e. the daughter isotope is A+1).

A variation on β-decay is electron capture decay. This time, a proton-rich nuclide 'hijacks' an electron from the inner electron shell transforming a proton into a neutron and emitting an electron neutrino. Electrons and electron neutrinos are fundamental particles known as first generation leptons and we can think of the electron neutrino as being to an electron what a neutron is to a proton (i.e. an uncharged electron). Electron capture destroys a proton so produces an isobaric daughter product with an atomic number one less than the parent isotope (i.e. A–1). A further variant of isobaric decay is a positron (a positively charged β-particle) decay in which a proton spontaneously converts into a neutron, and emits a positron and an electron neutrino. Whether electron capture or positron emission is favoured depends on the energy difference between parent and daughter isotopes. A small energy difference restricts the decay mode to electron capture whereas larger energy differences allow positron emission. Though the mechanisms differ, the net effect is identical in generating an A–1 daughter product.

In some cases nuclides can decay along different isobaric pathways: so-called branched decay. So, about nine out of ten potassium-40 atoms ($^{40}_{19}K$) decay by β-decay to calcium-40 ($^{40}_{20}Ca$), about 10 per cent decay by electron capture to Argon-40 ($^{40}_{18}Ar$), and occasionally (c.0.001 per cent) ^{40}K decays to ^{40}Ar by positron emission. The ratio of decays to Ca over decays to Ar is known as the branching ratio (Figure 3).

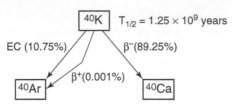

3. Branched decay of ^{40}K to ^{40}Ca by beta decay (β^-), and ^{40}Ar by electron capture (EC) and positron decay (β^+).

γ-decay

γ-decay is usually described as the emission of γ-rays but here we run into the quantum mechanics concept of wave–particle duality whereby matter is simultaneously composed of particles which have mass and also waves that transfer energy. γ-rays are extremely high frequency, high energy, and short wavelength electromagnetic radiation. Although the exact distinctions are blurred, γ-rays are essentially high energy X-rays. They are produced in association with other radioactive decays. Typically, α- and β-decays result in daughter nuclides which are in an 'excited' energy state. γ-decay reduces the nuclide to a lower energy state within a split second (usually less than a trillionth of a second) of an α- or β-decay. As such, γ-decay does not 'transmute' the nuclide into a different element but merely reduces the energy state of the nucleus of the daughter element. However, the energy stored in excited nuclear states is not insignificant. γ-rays are less ionizing than α- and β-particles, but that in turn makes them more penetrating, so that a thin metal foil shield that would not be penetrated by α-particles might be effectively transparent to γ-rays.

Spontaneous fission

The final form of radioactivity that we need to discuss is spontaneous fission. This is restricted to the heaviest elements, including the transuranic elements that occur in the actinide group with heavier nuclei than uranium. Of the natural elements only uranium and thorium are fissile, and the number of fissions is very small compared with other radioactive decays. For example, ^{235}U will undergo two billion decays for each fission. However, fission generates high energy, high mass atoms whose effects are easily recognized as tracks of damage in the lattices of U- and Th-rich minerals like zircon. Fission occurs because the heavy nuclei are inherently unstable and spontaneously divide into more energetically stable nuclides. A whole range of fission products are produced but cluster around masses 89–100 and 130–150. Some of the best-known fission products are ^{90}Sr and ^{137}Cs, which by virtue of their relatively long half-lives (twenty-nine and thirty years, respectively) are amongst the most persistent radiological contaminants in the aftermath of nuclear accidents. Note that the combined mass of the two peaks of fission products sums to slightly less than that of the fissile isotopes, ^{235}U in a conventional enriched-U nuclear reactor. This is because some mass is lost in the fission process through the loss of free neutrons and the conversion of mass into kinetic energy during fission.

Chapter 2
Isotopic clocks: the persistence of carbon

Radiocarbon

As we have already seen, about one in a trillion atoms of carbon is the radioactive isotope, ^{14}C. In the mid-1940s Willard Libby proposed that ^{14}C might be used to date organic material if the half-life of ^{14}C could be determined and the initial abundance of ^{14}C could be assumed. ^{14}C is a 'cosmogenic isotope' produced by the interaction between atmospheric nitrogen (N) and cosmic rays. Cosmic ray neutrons strike ^{14}N atoms which in turn produce ^{14}C and a free proton. The ^{14}C atoms rapidly combine into carbon dioxide, which quickly diffuses through the atmosphere and dissolves in the oceans.

At any one time the ^{14}C concentration of the atmosphere is constant. Photosynthesizing plants and respiring animals inherit that concentration and maintain equilibrium until they die, whereupon the ^{14}C stop-watch starts to tick. The half-life of ^{14}C is 5,730 years, so every 5,730 years the ratio of ^{14}C to ^{12}C will halve. With ultra-sensitive accelerator mass spectrometers (see Chapter 6) we can measure ^{14}C abundances over seven or eight half-lives before modern carbon contamination in the preparation laboratory overwhelms the overall signal. So, radiocarbon is used to date samples anywhere between the present day and about 50,000 years before present.

De Vries effect

Originally, it was assumed that atmospheric ^{14}C production would be a constant process and consequently that it was safe to assume that the initial ^{14}C/^{12}C had been similar to that of the present day over the whole range of the radiocarbon clock. However, this quickly proved to be a false assumption. As the quality of radiocarbon measurements improved, discrepancies emerged between the ^{14}C ages and the known dates of samples whose age could be constrained by other historical methods. In 1958, Hessel de Vries used wood samples of known age to demonstrate that atmospheric ^{14}C/^{12}C have varied through time (Figure 4). Two main effects influence ^{14}C production rates. First, fluctuations in the Earth's magnetic field regulate the flux of cosmic rays because the field partially shields the Earth from cosmic radiation. When the field is strong, fewer cosmic rays reach the atmosphere and ^{14}C production is low. Conversely a weak magnetic field encourages ^{14}C production because cosmic radiation is enhanced. Second, variations in sun spot activity influence the

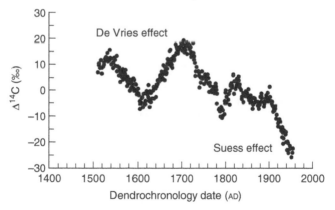

4. Variation in Δ^{14}C (i.e. deviation in ^{14}C/^{12}C from a standard value) through time, recording natural variations in ^{14}C production (de Vries effect) and the rapid drop in atmospheric ^{14}C from burning of fossil fuels in the industrial era since about 1900 AD (Suess effect).

primary production of cosmic rays which in turn controls ^{14}C production rates.

This 'de Vries effect' became well understood by comparing ^{14}C/^{12}C in samples of trees which could be independently dated using their annual growth rings. The bristlecone pines and giant sequoia trees of California and elsewhere offer tree-ring records (dendrochronology) that extend back over 5,000 years. Beyond this, dendrochronologists have been able to extend records using fossil trees and matching the annual rings where they overlap with the living records and with one another. Beyond the tree record, fossil corals, which can be independently dated using isotopes of uranium and thorium, have been used to calibrate the de Vries effect throughout the 50,000 year range of radiocarbon dating.

Suess effect

If we examine the variation in atmospheric ^{14}C production rate two further phenomena can be identified. First, there is a major reduction in ^{14}C between about 1850 and 1950, which was first recognized by Austrian chemist Hans Suess. This Dr Suess realized that the dilution of atmospheric ^{14}C coincided with the onset of the industrial revolution and the attendant massive increase in burning of fossil fuels. Oil, gas, and coal require millions of years of geological time to form and mature. For example, most of the coal in Britain and the USA formed in the Carboniferous period (the clue is in the name) of geological time, which spanned roughly 360 to 300 million years ago. Even one million years is about 175 ^{14}C half-lives. So, any original ^{14}C in fossil fuels decayed away many millions of years ago. Thus, burning oil, gas, and coal introduces ^{12}C and ^{13}C to the atmosphere (as CO_2) but no ^{14}C, and the result is the reduction in atmospheric ^{14}C that we now call the 'Suess effect' in recognition of its discoverer. In making this observation, Hans Suess was among the first to recognize that taking millions of years' worth of stored carbon and releasing it very quickly, in geological terms, was changing the composition of Earth's atmosphere. In essence, Suess laid the foundations for the scientific consensus that fossil fuel use has greatly influenced the global climate.

Willard Libby received the Nobel Prize in Chemistry in 1960 for the discovery of radiocarbon dating. Hessel de Vries was mentioned only briefly in Libby's Nobel acceptance speech, though many consider de Vries's contribution to making radiocarbon dating work warranted a share of the Nobel. However, the previous year, de Vries had died. In an apparent act of unrequited love, de Vries murdered Anneke Hoogeveen, who had been an analyst in his laboratory who had become engaged to another man. After murdering Anneke, de Vries took his own life. Surely the mores of the early 1960s led Libby to gloss over de Vries's contribution to the radiocarbon revolution in archaeology.

Bomb carbon

The second obvious trend in atmospheric ^{14}C production is a major spike through the 1950s that peaked about 1965 (Figure 5). This was as a result of nuclear weapons testing, which produced a huge supply of neutrons that interacted with stable carbon atoms to double roughly the ^{14}C concentration in the atmosphere. Here again is an example of how easily humankind has been able to disrupt the natural system of the Earth's atmosphere. Since 1965 this 'bomb spike' has begun to decay following the partial test ban treaty of 1963. So, in the second half of the 20th century, the atmospheric ^{14}C concentration changed much more rapidly than would have been expected from natural variations. This has enabled a second type of radiocarbon dating to be applicable from the mid-20th century to the present day. In this method, dating depends not on the radioactive decay of ^{14}C but simply on the fact that organic material inherits the $^{14}C/^{12}C$ of the atmosphere at the time of death. So, in the nuclear age,

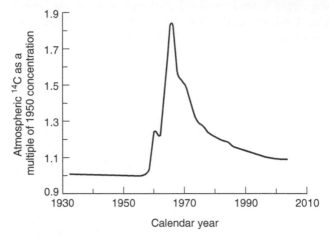

5. ^{14}C associated with atmospheric nuclear tests and subsequent decay towards natural abundance.

comparing that ^{14}C/^{12}C with the atmospheric ^{14}C curve gives the age of death—and the changes in atmospheric ^{14}C have been so rapid that death can be determined to within one or two years—much more precise than with conventional radiocarbon dating.

Increasingly, bomb carbon dating is being used as a forensic tool, and it is not uncommon these days in the foyer of the institution I direct to encounter police officers brandishing human bones for ^{14}C analysis. In 2008, allegations of historical child abuse led to a police search of the Haute de la Garenne in Jersey. The building, which was being used as a youth hostel by 2008, had previously been a children's home and several bone fragments were recovered by the police. The discoveries became front-page news and generated huge amounts of press speculation. Subsequent trials have revealed that child abuse did occur at Haute de la Garenne. However, of a large quantity of bone fragments recovered by the police, only three turned out to be human, and radiocarbon dating was able to demonstrate that none of these were modern.

Under the Convention on International Trade in Endangered Species (CITES) and UK and EU law it is illegal to trade in ivory, but there is an exception for 'antique' specimens, with antique being defined as pre-dating 1st June 1947. Since the nuclear age began with the Hiroshima and Nagasaki bombings in 1945, all illegal ivory will contain bomb carbon. Thus, it is straightforward to distinguish legal and illegal ivory trading simply by measuring its ^{14}C concentration.

Similarly, bomb radiocarbon has been used to check the authenticity of historical Scotch whisky. A genuine pre-1945 spirit will have the natural abundance of ^{14}C of the time at which its barley was harvested, whereas a modern fake would be contaminated with bomb carbon—readily distinguished by a relatively simple radiocarbon measurement.

Actually, we have now diversified into using ^{14}C as a source tracer instead of as a simple chronometer. Here, we return to the simple fact that geological sources of carbon are ^{14}C-free whereas sources that were alive in the past few tens of thousands of years retain radiocarbon. So, for example, if we take a sample of atmospheric soot, say from an air filter collected in a city centre, its ^{14}C could tell us the extent to which it derives from fossil fuel burning (dead carbon) or burning of biomass (live carbon). Equally, scientists are using a similar approach to check that mixed fuels contain the correct proportions of fossil fuel (dead carbon) and biofuel (live carbon).

Irrespective of the relative contributions, Libby, de Vries, Suess, and their contemporaries established a dating technique that has become ubiquitous. These days, many thousands of radiocarbon measurements are made across the world every year with individual laboratories responsible for a few thousand ^{14}C dates and tracer measurements per year. Without doubt radiocarbon has become one of the most valuable isotopic tools available to scientists looking to date materials formed over the past 50,000 years, or to trace and apportion different sources of carbon. With increasing consensus that current-level human CO_2 emissions are unsustainable without significant climate change, one can only credit the radiocarbon pioneers with providing a scientific technique that has had, and will continue to have, major impact on a scientific endeavour of enormous societal relevance.

Chapter 3

You are what you eat...plus a few per mil

Delta notation

The rarer stable isotopes (e.g. ^{13}C, ^{15}N, and ^{18}O) tend to constitute only 1 per cent or so of the natural element, so ratios of ^{13}C to ^{12}C, ^{15}N to ^{14}N, and ^{18}O to ^{16}O will always be small numbers. To avoid strings of zeros after the decimal point, isotope geochemists tend to use delta (δ) or 'del' notation. This simply expresses the isotopic composition of a sample relative to that of some standard material multiplied by 1,000 to generate easy-to-handle whole numbers. So, for carbon, δ^{13}C is defined as:

$$\delta^{13}C = \left[\left(^{13}C/^{12}C_{\text{sample}} - ^{13}C/^{12}C_{\text{standard}} \right) / ^{13}C/^{12}C_{\text{standard}} \right] \times 1,000.$$

which can be simplified to:

$$\delta^{13}C = \left[\left(^{13}C/^{12}C_{\text{sample}} / ^{13}C/^{12}C_{\text{standard}} \right) - 1 \right] \times 1,000.$$

In multiplying by 1,000 we generate a per mil (or per mille: ‰) value which is entirely analogous to a percentage but 1‰ = 0.1 per cent.

For carbon, the standard was originally 'PDB', a fossil belemnite taken from the Cretaceous Pee Dee formation from South Carolina. Nitrogen isotopes are standardized to the isotopic composition of air (AIR) and oxygen was referenced to standard mean ocean water (SMOW). These days it is more common to see C and O isotope ratios referenced to V-PDB and V-SMOW, where V stands for Vienna, the headquarters of the International Atomic Energy Agency (IAEA), which promulgated the worldwide use of common reference standards in stable isotope measurements. In practice, isotope laboratories maintain their own reference gases which are carefully calibrated against the IAEA standards.

Organic isotopic fractionation

Isotopic fractionation is the effect in which any process changes the relative abundance of the isotopes. So, a mineral crystallizing from a fluid at low temperature would usually have a different isotopic composition from the fluid in which it grew. The difference between mineral and fluid isotopic composition is called fractionation. Overall, an isotopic balance has to be maintained, so the isotopic fractionation into the mineral will be balanced by a change in the isotopic composition of the remaining fluid. In practice, the fluid reservoir is often large compared to the amount of mineral crystallized, so removing a small amount of mineral of highly fractionated isotopic composition will be balanced by only a small change in the isotopic composition of the semi-infinite fluid.

When plants and animals generate tissues, they also have to preserve an isotopic balance. However, what goes in doesn't necessarily come out, because organisms often fractionate isotopes, retaining one isotope preferentially in the cells produced and expiring CO_2 depleted in that isotope. In animals, the whole body $\delta^{13}C$ is a reflection of the dietary $\delta^{13}C$, but animals tend to retain ^{13}C in their body tissues by about 1‰ and expire CO_2 that is about 1‰ depleted in ^{13}C. So, isotopically you are what you eat plus a small (few per mil) isotopic fractionation. In detail, the processes are complicated, and different tissues and different organic molecules preserve different isotopic signatures.

Plants are a little more complicated because one of the major controls on carbon isotope distribution in the natural environment is photosynthesis: the process by which plants and other organisms turn sunlight into chemical energy to fuel themselves. Different mechanisms of photosynthesis have evolved and they have very different effects on C isotope distribution. So-called 'C3' plants use the enzyme RuBisCo in their leaves to fix atmospheric CO_2. C3 fixation can be achieved at moderate temperature, with moderate sunlight, and in the presence of plentiful water. However, C3 fixation is relatively inefficient and imparts a large shift in $\delta^{13}C$. C3 plants have $\delta^{13}C$ of about −24 to −34‰. C4 plants have evolved a more efficient mechanism for delivering CO_2 to RuBisCo through modifications in leaf anatomy. While C4 fixation places greater demands on environmental conditions, it makes more efficient use of CO_2 and causes less C isotope fractionation. C4 plants have $\delta^{13}C$ of about −6 to −19‰.

Millet and maize are C4 plants whereas rice, wheat soya, and potatoes are C3. Using the principle of 'you are what you eat plus (only) a few per mil' it ought to be possible to reconstruct the staple diet of fossil animals assuming that those fossils preserve the fidelity of the isotopic record. In practice, bone collagen, chitin, and the organic fraction of shell material do appear to preserve the *in vivo* stable isotope signal.

Native North American skeletons recovered at archaeological sites were analysed for $\delta^{13}C$. Earlier than about 1,000 years before present, the bones have $\delta^{13}C$ of about −20‰ showing a dominance of C3 plants in their diets. However, after about 1000 AD, the carbon isotopes shift to range between about −15 and −10‰: a clear C4 signature. These data are interpreted to indicate the onset of maize cultivation in North America about 1,000 years ago.

Nitrogen isotopes are also fractionated by organisms (Figure 6). Typically, tissues are enriched in ^{15}N and balanced by excretion of,

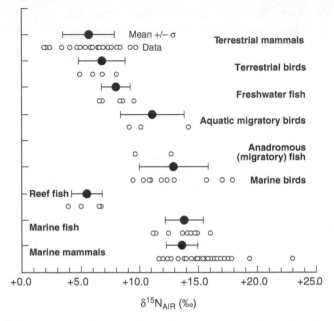

6. Nitrogen isotope composition (relative to that of an air standard) of the bone collagen of a variety of animals.

for example, urine depleted in [15]N. Actually, this ^{15}N-enrichment is remarkably consistent with about a 3‰ [15]N enrichment at every (trophic) level in the food chain. Thus, herbivorous animals will inherit the $\delta^{15}N$ of the plants they eat but lay down tissues that are 3‰ higher $\delta^{15}N$. The carnivores that prey on the herbivores will have whole-body $\delta^{15}N$ a further 3‰ enriched in [15]N. So there is a progressive increase in $\delta^{15}N$ moving to higher trophic levels in the food web. Such information is of great value to ecologists in understanding food chains and how they have been affected by environmental conditions—particularly how environmental and climatic change has impacted the nutritional behaviour of animals.

In addition to the trophic level effect, which can be discerned in both terrestrial and marine animals, there is also a strong N-isotope discrimination between terrestrial and marine animals that can be discerned by N-isotope measurements of bone collagen. Marine mammals, fish, and birds all show higher $\delta^{15}N$ than their land-based counterparts, with very little overlap. There are, however, some notable but explicable exceptions. Aquatic migratory birds have $\delta^{15}N$ values that are intermediate between fully marine and fully terrestrial species, reflecting the fact that they feed in both marine and terrestrial environments. Similarly, fish that migrate between terrestrial and marine habitats (e.g. salmon) have intermediate $\delta^{15}N$. Terrestrial fish that are carnivorous for most of their life-cycle have slightly higher $\delta^{15}N$ than terrestrial herbivores reflecting their feeding higher in the food chain.

There is one group of animals, reef-dwelling fish, whose bone collagen $\delta^{15}N$ values are inconsistent with their marine diet. Such fish have low $\delta^{15}N$ well within the terrestrial range and anomalously high $\delta^{13}C$. The nitrogen isotope signature probably reflects the large amount of nitrogen fixation by blue-green algae around coral reefs compared to the open oceans. Similarly, the high $\delta^{13}C$ signal most likely reflects feeding on sea-grass and coral zooids (the living element of corals) which themselves concentrate ^{13}C.

King Richard III

An interesting application of stable isotopes in reconstruction of diet concerns the radiocarbon dating of the skeleton discovered beneath a car park in the UK city of Leicester. Initial radiocarbon dating by scientists at the Scottish Universities Environmental Research Centre yielded ages of 1430–60 AD and at Oxford University gave 1412–49 AD, both significantly older than 1485 when King Richard III died at the battle of Bosworth Field. So, it seemed unlikely that these bones were those of the King, despite displaying the severe curvature of the spine for which Richard is famous. However, stable isotope analysis suggested that this skeleton belonged to someone whose diet contained a substantial contribution from seafood—which in itself was an indication that this was a high-status individual. As well as characteristic $\delta^{13}C$ and $\delta^{15}N$ isotope values, marine organisms are also depleted in ^{14}C. Because of the way carbon is cycled through the marine environment, seawater and animals that live in the sea incorporate 'dead' carbon and give radiocarbon ages that are up to about 400 years older than the date of death. Using the stable isotopes to correct this 'marine reservoir effect' shifted the radiocarbon age of the skeleton to between 1450 and 1540 AD, bang on the age expected for King Richard III. Subsequent DNA analysis proved a genetic link to the King's decedents making a strong case that the skeleton is indeed that of Shakespeare's 'son of York'.

Geography and isotopic composition

During evaporation of water, the evaporating water vapour is enriched in ^{16}O because $H_2^{16}O$ has a slightly higher vapour pressure than $H_2^{18}O$. The flip-side of this phenomenon is that snow and rain condensing from clouds are richer in ^{18}O than the clouds themselves. So, water vapour in the atmosphere has $\delta^{18}O$ about 13‰ lighter than the seawater from which it evaporated (i.e. $\delta^{18}O$ = −13‰ from $\delta^{18}O$ = 0‰ seawater). The first rain or snow from that air-mass is about 10‰ heavier (−3‰) and removal of that water drives the remaining air-mass to even more negative $\delta^{18}O$. As air moves across a continent progressively dumping water as rain and snow its oxygen isotope ratio evolves (Figure 7).

Variation in the intensity of evaporation means that the $\delta^{18}O$ of precipitation is geographically zoned with a tendency for low $\delta^{18}O$ at high (low evaporation) latitudes and progressively ^{18}O-rich rainfall towards the (high evaporation) equator. For example, in North America there is a continuous zonation north-west to south-east from $\delta^{18}O$ $c.$−30‰ in Alaska to $c.$−4‰ in Florida. So,

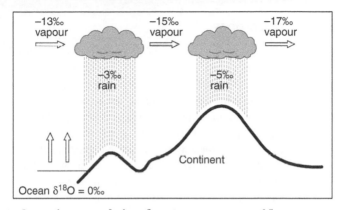

7. Oxygen isotope evolution of a water mass evaporated from seawater and progressively transported across a hypothetical continent.

if we understand (and can correct away) metabolic fractionation of oxygen isotopes in an animal, we should be able to reconstruct the $\delta^{18}O$ of its drinking water by analysing the O isotope compositions of the biominerals that construct the animal's bones. Someone from Alaska could easily be distinguished from someone from Florida. Indeed, if suitable samples could be identified, one would expect that Lou Reed's semi-fictional Holly in the song 'Walk on the Wild Side' who hitch-hiked from Miami to join Andy Warhol's entourage in New York City would have recorded the journey in the $\delta^{18}O$ of the bones (s)he grew along the way.

This geographical O isotope zonation is disrupted by major changes in topography. As an air-mass rises over elevated topography, the associated cooling produces precipitation which is enriched in ^{18}O, leaving behind low $\delta^{18}O$ water vapour. The upshot is that areas of high topography have anomalously low $\delta^{18}O$ for their latitude. The effect is worth about 0.2‰ per 100 metres (m) below about 3,000 m, rising to 0.5‰ per 100 m above 3,000 m. For example, in the Bernese Alps precipitation at 500 m has $\delta^{18}O$ of about –7‰ whereas it is down to –17‰ at 4,000 m.

23

Strontium isotopes

Strontium (Sr) is one of the alkaline earth metals located directly below its more abundant neighbour calcium in the periodic table (Figure 2). Its chemical similarity to calcium means that Sr readily substitutes for Ca in body materials like bones and teeth. The source of Sr in animals and plants will be drinking water and Ca-rich foods like dairy products. Ultimately though, the Sr in the diet comes from the rocks and minerals that break down to make the soils that plants grow on, which cattle then eat and turn into the milk that is such an important source of calcium in the Western human diet. It is worth noting that diets that are less dairy-based and take much of their calcium from seafood, e.g. Japanese, will be dominated by marine rather than terrestrial Sr and this should be recorded in the isotopic composition of Sr.

Sr has four natural isotopes—84, 86, 87, and 88—all of which are stable. In addition, there are radioactive isotopes, 89 and 90, which are well-known as dangerous fission products associated with nuclear explosions and accidents. Of course, it is the propensity of Sr to enter bones that makes us so vulnerable to radiostrontium. The relative mass difference of the Sr isotopes is small (e.g. the difference between 87 and 86 is $1/86 \times 100 = 1.16$ per cent compared to that between ^{16}O and ^{18}O which is $2/16 \times 100 = 12.5$ per cent). This means that the amount of natural fractionation of Sr isotopes is expected to be negligible. In fact, some of the most effective mechanisms that do fractionate Sr isotopes are the very processes that we use to ionize Sr in the mass spectrometers used to measure Sr isotope compositions. Thus, in order to make accurate Sr isotope ratio measurements we have to correct away fractionation induced during the measurement, and this also removes any record of natural fractionation.

^{84}Sr, ^{86}Sr, and ^{88}Sr are 'primordial': they have been in the Solar System since it formed. ^{87}Sr is about 90 per cent primordial but the remaining 10 per cent was produced by the β-decay of ^{87}Rb through geological time. The $^{87}Sr/^{86}Sr$ of the whole Earth when it formed was slightly less than 0.700. Today it is closer to 0.705 but the Earth is divided into the continental crust with high $^{87}Sr/^{86}Sr$ (0.720) and the Earth's deep mantle with low $^{87}Sr/^{86}Sr$ (0.703). This zonation reflects the behaviour of Rb during the formation of continental crust. Because the crust has high Rb/Sr it has evolved high $^{87}Sr/^{86}Sr$. By contrast the Earth's mantle is relatively depleted in Rb, has low Rb/Sr, and therefore evolves $^{87}Sr/^{86}Sr$ more gradually.

For any rock, its $^{87}Sr/^{86}Sr$ is a combination of the $^{87}Sr/^{86}Sr$ it had when it formed, its Rb/Sr and the length of time since its formation, during which ^{87}Rb has decayed into new ^{87}Sr. For our present purposes it is sufficient to understand that different geological formations will have different ^{87}Sr and often it will be possible to characterize individual locations by the Sr isotope composition of its rocks. It is this geological Sr isotope signature that animals inherit when they lay down Ca-bearing tissues.

Different body parts can represent different periods in the lifetime of an animal. For example, human tooth enamel is, obviously, formed in childhood. By contrast, bones carry a more complicated record because bone is continually 'remodelled'. In childhood, this bone turnover is extreme—in the first year of life our bodies replace our entire skeleton. Later in life this reduces, but it is still about 10 per cent per year in a mature adult. Failure of this process is at the heart of conditions like osteoporosis. Thus, a Sr isotope record from a human skeleton can be considered to be some sort of ten-year average of the geographical movements of that person.

Ötzi the Iceman

A fascinating case study of the use of O and Sr isotopes in an archaeological context is the reconstruction of the life history of 'Ötzi the Iceman'. Ötzi was discovered in September 1991 by two walkers who had strayed off the well-trodden path in the Ötztal

Alps near the border between Austria and Italy at a height of 3,210 m. At first, it was assumed the body was that of a recently deceased climber, but it soon became clear that this was a far more ancient corpse preserved in ice and naturally mummified. We now know that Ötzi died in his mid-40s about 5,200 years ago.

The corpse's DNA is only partially preserved and indicates similarities with central and northern Europeans. While DNA analysis has advanced greatly in recent years, when Ötzi was found, the nuclear DNA, which could offer better spatial resolution, was not sufficiently well preserved. However, the pollen and moss recovered from Ötzi's intestines suggest residence in northern Italy shortly before his death. Could isotopic analyses reveal more about his origins and lifestyle?

As it happens, the corpse was discovered at an Alpine watershed with considerable topographic contrast. To the north of the watershed, precipitation is mainly derived from the Atlantic and has been transported considerable distances. To the south, the precipitation is mainly Mediterranean-derived and close to its source. Thus, we expect northern precipitation to have lower $\delta^{18}O$ compared to southern rain, and this was confirmed by analysis of contemporary rivers. Ötzi's tooth enamel has $\delta^{18}O$, which indicates drinking water with $\delta^{18}O$ between −10.6 and −11.0‰. Such values are only consistent with a southern source and suggest that he lived in what is now northern Italy when he was growing his second teeth.

Sr isotope measurements on tooth enamel immediately rule out a number of areas as Ötzi's childhood home, and the data are pretty definitive that he grew up in an area characterized by metamorphic rocks called gneisses (the 'g' in gneiss is silent). Adding lead isotopes produced a closest match between Ötzi's tooth enamel and local geology at a place called Feldthurns. Independent archaeological evidence indicates a copper age settlement at Feldthurns, so it seems quite likely this was Ötzi's childhood home.

Ötzi's bones indicate that during adulthood his drinking water had $\delta^{18}O$ between −11.4 and −11.7‰ suggesting that he moved from his childhood home but still lived to the south of the discovery site. Bone $^{87}Sr/^{86}Sr$ values are lower than the teeth values and further support movement into an area around present-day Merano in the Etsch valley. However, there is another possible scenario, Ötzi's lifestyle may not have involved permanent relocation but could have been semi-nomadic. If Ötzi was involved in transhumance agriculture, then the decrease in drinking water $\delta^{18}O$ between childhood and adult life would be consistent with adult summers spent at altitude.

Whatever Ötzi's business in the mountains, he seems to have met with a violent end. Although initially elusive, a chest X-ray revealed a flint axe head lodged in his left shoulder. An unhealed wound to his hand suggests that Ötzi was in a fight hours or days before his death. There was also a serious blow to his head and a fracture of his skull which may indicate a blow or a fall. Somebody had it in for Ötzi, and isotopes have helped to provide background to the life of this 5,000-year-old murder victim.

Chapter 4
Measuring isotopes: counting the atoms

We have now seen some of the things that we can do with isotopes but, until now, we have assumed that it is possible to know the abundances of isotopes in a variety of materials. Of course, such knowledge depends on the ingenuity of analytical chemists who have figured out ways of measuring isotope abundances. Let's take some time to explore how isotopes are measured. Whether isotopes are stable or radioactive primarily determines how they can be detected. For stable isotopes there is only one choice: we have literally to count the number of atoms of each isotope using an instrument called a mass spectrometer. For radioactive isotopes there are two options; we could count the atoms (i.e. mass spectrometry) but, for short-lived isotopes, their low abundance often makes them difficult to detect. So, the alternative is to use their radioactivity as a give-away to their presence. In these nuclear spectroscopy or counting methods, we detect the characteristic energy released by the decay of a particular isotope. However, nuclear spectroscopy can be very time consuming because we have to wait for radioactive decays to happen naturally over days and weeks. By contrast, mass spectrometry turns the atoms into charged ions and separates them into the different isotopic species using some type of mass filter. Mass spectrometers can ionize millions of atoms allowing precise measurements of isotopic abundance over a few minutes or hours.

Accuracy and precision

Before going any further, it is useful to introduce a couple of concepts that are meat and drink to the analytical chemist: accuracy and precision. In everyday life these are roughly synonymous but for analysts they have a subtle, but very important, different meaning. I'll try to explain what we mean using the shooting targets illustrated in Figure 8. The first target shows bullet holes peppered over the target. Assuming the object is to hit the bull's-eye, the shooter of this target was both inaccurate and imprecise—he repeatedly missed the bull's-eye with very little consistency. The second target shows a gun-man who is inaccurate but precise; he missed the bull's-eye but hit the same part of the target with each shot—maybe he is a great shot but has a gun whose telescopic sight needs to be realigned? The third target shows bullet holes that are both accurate and precise. The shots hit the bull's-eye (accurate) and are closely grouped (precise). So, in analytical chemistry, getting the right answer is being accurate. Getting the same answer time after time is being precise. Being precise is not a guarantee of accuracy but, ironically, it is easier to be accurate if your measurement is imprecise.

In a perfect analytical instrument designed to measure isotope abundances, precision is limited by what we call 'counting

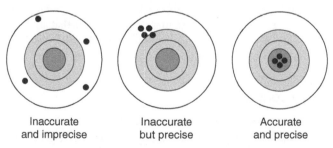

Inaccurate and imprecise	Inaccurate but precise	Accurate and precise

8. **Illustration of the concepts of accuracy and precision.**

statistics'. Assuming everything works perfectly, the precision of a measurement is determined by the number of atoms measured. You can't get better than the ratio of the square root of the number of measurements to the total number of measurements, i.e.:

$$\text{Percentage Precision} = \sqrt{N}/N \times 100$$

So, if we measure 10,000 atoms, we get $\sqrt{10{,}000}$ (100)/10,000 × 100 = 1 per cent precision. Let's see how that impacts on nuclear spectroscopy. A mole of any element has about 6×10^{23} atoms (i.e. Avogadro's number). A mole of carbon weighs 12 grammes, so a milligramme is about 5×10^{19} atoms. Of these, only one in a trillion is ^{14}C, i.e. 5×10^7. The half-life of ^{14}C is 5,730 years, so in that time 2.5×10^7 atoms will decay (i.e. about 4,400 decays per year). So, even if we can design an instrument with 100 per cent counting efficiency, it will still take over two years of counting to detect the 10,000 decays we need to get 1 per cent precision from our perfect instrument.

By contrast, let's assume we can design a mass spectrometer that is capable of separating ^{14}C from the other carbon isotopes and other interferences (e.g. ^{14}N), which we will see later is not a trivial task. Then we'll assume that instrument is capable of generating ion beams of 1μA, that's 6.25 trillion carbon ions per second, of which 6.25 per second will be ^{14}C, so to measure 10,000 ^{14}C atoms will take 10,000/6.25 = 1,538 seconds or about twenty-five minutes. So, the challenge has been to design mass spectrometers that are capable of detecting very rare isotopes while retaining the nuclear spectroscopy techniques for very short-lived isotopes.

Geiger counter

The best-known and most ubiquitous radiation detector is the Geiger counter (Figure 9) which is a combination of a Geiger–Müller (GM) tube and processing and display electronics. The GM tube consists of a central wire electrode carrying a positive electric

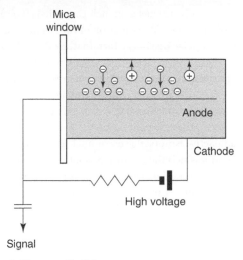

9. Schematic diagram of a Geiger counter.

charge (anode) while the wall of the tube is negatively charged (cathode). The tube is filled with a low pressure inert gas (He, Ne, or Ar). When a radioactive particle enters the GM tube, molecules of the gas are ionized, creating positively charged ions and electrons (ion pairs), which are attracted to the cathode and anode, respectively. Close to the anode the electrons gain enough energy to ionize further gas molecules and create an 'avalanche' of electrons (Townsend avalanche) which amplifies the signal producing a significant output from a single ionizing event. While the Geiger counter is able to detect radioactivity efficiently (especially higher energy α- and β-radiation) it does not offer any ability to resolve different energies of events—which would offer spectroscopic capability.

Gas proportional counter

The gas proportional counter is a modification of the Geiger counter that does offer some spectroscopic ability. In this detector, the inert gas (usually Ar) has a quench gas added (typically 10 per cent of methane). The function of the quench gas is to ensure that each pulse discharge terminates within the gas. Rather than the multiple electron avalanches of the GM tube, the gas proportional counter carefully ensures that only a single avalanche occurs. Under such conditions the number of ion pairs generated is proportional to the energy of the initial radioactive decay, thereby generating the ability to discriminate different radioactive isotope decays or nuclear spectroscopy. The conditions under which the gas proportional counter functions correctly require careful design to optimize the voltage, geometry, and diameter of the anode to ensure the single avalanche condition.

Scintillation counters

Scintillation detectors exploit the effect that certain solids and liquids react to the absorption of energy from radioactive decay by emitting energy in the form of light. Essentially, the radiation causes electrons to be excited to quantum levels above the lowest energy ground state. As those electrons return to the ground state the loss of energy is transformed into light in the process we call scintillation. The quantity of light emitted is proportional to the radiation energy and we have the basis for distinguishing different energies which constitutes spectroscopy.

In liquid scintillation spectrometry, the scintillating material is known as the scintillation cocktail. Basically, the cocktail is a mixture of an aromatic solvent into which the sample material will dissolve and a scintillating solute or fluor which generates the pulses of light. Samples are introduced to the detector in vials made of low-potassium borosilicate glass or high-density polyethylene and can range from 'raw' samples, for example for ^3H analysis in environmental water samples, through to complex pre-concentrations of elements extracted from environmental samples (e.g. ^{90}Sr in milk, soil, vegetation, and urine).

When a radioactive decay occurs, the α- or β-particle (liquid scintillation is not efficient for most γ-ray detection) interacts with the aromatic solvent which is far more abundant than the fluor. The solvent accepts the energy lost by the ionizing particle and becomes activated. The excess energy is then transferred to the fluor molecule, which dissipates the energy by emission of a photon of light. The intensity of the flash of light is a function of the energy of the decay.

To detect the light photons we use a photomultiplier tube (PMT), which is the same technology used in image intensifying cameras used by wildlife photographers to film nocturnal animals or the night-sights used by police marksmen. A photon hitting the PMT encounters a photo-cathode made of a combination of elements (e.g. antimony-rubidium-caesium) that demonstrate the photoelectric effect in which the incident photons release electrons from the cathode. Enclosed in a glass tube under vacuum, the electrons produced are attracted by an ever-increasing positive voltage towards a series of plates called dynodes that are made of a material (e.g. copper-beryllium alloy) that generates a burst of electrons for every incident electron. The electrons cascade down the PMT and a greatly amplified pulse is detected at the ultimate anode. Let's assume we have a PMT with fifteen dynodes and the final anode, and that every dynode generates ten electrons for every incident electron, that's 10^{15} electrons arriving at the anode for every incident photon. 10^{15} electrons have a cumulative charge of 1.6×10^{-3} coulombs. So, one photon per second generates a current of 1.6 milliamps which is readily amplified. The amplified signal is then converted to a digital signal using the same type of device (an analogue to digital converter or ADC) that you could use to convert vinyl records into MP3 files to listen to on your computer or iPod. Finally, a multi-channel analyser is used to sort the photons into their different original energies that are characteristic to the original radioactive decay energies that provide spectroscopic recognition of different radioactive isotopes.

In practice, PMTs have their own inherent noise which we need to eliminate to identify real radioactive decays. So, liquid scintillation detectors employ two PMTs. It is very unlikely that noise will occur in both detectors simultaneously because the cascade time through the PMT is very short, maybe twenty nano-seconds (i.e. 20×10^{-9} s). So, noise in either PMT will only be recorded by that PMT whereas real decays will be recorded by both PMTs—we call this a 'coincidence detector'.

In addition to PMT noise there will be natural radioactivity emanating from the laboratory environment that doesn't come from the sample under analysis. For example, potassium is a common element making up about 2 per cent of Earth's crust, so there will be potassium (K) in the laboratory, on the shoes of the lab workers, in the chemicals used to prepare samples, etc. ^{40}K is a radioactive isotope and we need to make sure that we don't confuse this radioactivity with that inherent to the sample we are trying to characterize. Additionally, cosmic rays which continually bombard the Earth from outer space are also capable of generating spurious readings in our detectors. The cosmic rays that generate our cosmogenic isotope chronometers (Chapter 8) are a nuisance in the nuclear spectroscopy laboratory and we need to exclude them from our radiation detectors by shielding.

Passive shielding is achieved by surrounding the detector with elements that soak up radioactivity. It is not unusual for liquid scintillation detectors to incorporate a tonne of lead shielding as well as cadmium and copper to absorb secondary X-rays and thermal neutrons. This will remove the environmental radioactivity associated with building materials and instrument construction components and the 'softer' cosmic radiation. However, 'hard' cosmic rays can even pass through the lead shield. To remove this interference we use active shielding which involves placing another radiation detector on the outside of the coincidence detector. Events detected by this 'anti-coincidence' detector are eliminated, so any event detected simultaneously by the coincidence and anti-coincidence systems originated outside the sample and can be subtracted from the signal attributed to the sample itself.

With effective shielding, liquid scintillation counting has become a favoured method for a variety of isotopic methods including radiocarbon (^{14}C) dating (though we will shortly see that this has been largely superseded by accelerator mass spectrometry), tritium (^{3}He) analysis of groundwater and other environmental samples, measurement of the shorter lived members of natural decay series (see Chapter 9) in environmental samples (e.g. ^{234}Th, ^{231}Pa, and ^{222}Rn), environmental monitoring associated with nuclear establishments to detect β-emitting isotopes which do not have significant γ-activity (e.g. ^{3}He, ^{14}C, ^{35}S, ^{90}Sr, ^{99}Tc, and ^{241}Pu), and measurement of gross α-activity in air filters and surface wipes during nuclear decommissioning.

Solid state detectors

Thallium-doped sodium iodide (NaI(Tl)) detectors are solid devices but follow the same basic principles as liquid scintillators. NaI(Tl) detectors are used in the laboratory environment when high counting efficiency is required but the mix of radioactive isotopes to be measured does not demand high spectral resolution. However, NaI(Tl) has the advantage that it can be manufactured as large single crystals around which γ-detectors can be constructed that are sufficiently robust to be taken out of the laboratory into field situations. NaI(Tl) is a highly efficient scintillation material, although it has limited ability to resolve different sources of γ-activity and careful separation of complex spectra is required in post-processing of the collected data. Nevertheless, useful field-deployable detectors range from small units that can be carried in back-packs and walked around field sites to larger systems that are still sufficiently portable to be mounted on vehicles or flown on helicopters or unmanned airborne vehicles. Such airborne systems have been used to establish the distribution of natural environmental radioactivity. Moreover, these are critical tools to establish the distribution of radioactive isotopes around nuclear facilities—especially in the aftermath of nuclear accidents.

A further class of radiation detectors consists of devices based on semi-conducting materials. We can think of the electrons in solids falling into two bands (Figure 10). Valence band electrons are bound to individual atoms whereas conduction band electrons are able to move around the crystal lattice. When electrical current flows through a wire it is the conduction band electrons that carry the electrical charge. In conductors, the conduction band is continually occupied and there is an energy continuum between valence and conduction bands. In insulators, the conduction band is permanently devoid of electrons. In semi-conductors there is an energy 'band gap' between the valence and conduction bands. The band gap for semi-conductors is sufficiently small that changes in temperature alter the population of the conduction band. At room temperature it is occupied with electrons and the material conducts. However, at very low temperatures the electrons fall into the valence band and the material becomes insulating.

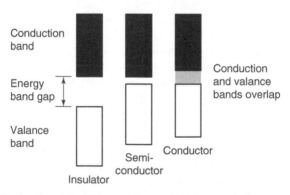

10. **Illustration of the difference between insulators, conductors, and semi-conductors.**

If we take a large crystal of germanium, which is an intrinsic semi-conductor, and apply a voltage to it while holding it at very low temperature no current will flow because the conduction band is empty. However, if a γ-ray loses its energy in the germanium there is enough energy to excite electrons into the conduction band and the result is an electrical pulse. Moreover, the pulse size is proportional to the radiation energy. If we amplify the pulse while maintaining the proportionality to the original signal, convert it to a digital signal, and analyse the signal with a multi-channel analyser, we have an effective γ-spectrometer (Figure 11).

γ-rays cause relatively little ionization as they pass through materials compared to α- and β-particles. This is why γ-sources need to be enclosed in thick lead shields while pure α-emitters can be contained by as little as a sheet of paper. α-particles do a lot of damage but don't travel very far; γ-rays do less damage but travel much further. In γ-spectrometry this is a good thing because it means there is very little loss of γ-radiation within the sample itself—so-called 'self-absorption'. This allows us to present

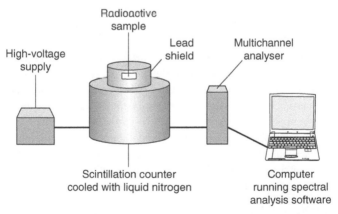

11. Diagram of a semi-conductor counting system for gamma emission spectrometry.

samples for γ-spectrometry as liquids or solids with minimal pre-treatment. Sediment samples, for example, are typically simply pressed into plastic containers for measurement. However, there are some important principles to be upheld for successful analysis. In particular, the geometry of the sample relative to the detector should be suitable and constant from sample to sample. Also, it is important to calibrate the efficiency of the detector for the type of sample being analysed, which requires careful matching of samples and standards.

Intrinsic semi-conductors, like the germanium detectors described already, need to be extremely pure, with very low levels of impurity. However, there are also so-called 'non-intrinsic' semi-conductors to which impurities have been deliberately added. These fall into two groups called N- and P-types for negative and positive. N-type semi-conductors (e.g. silicon or germanium) are deliberately doped with an element (e.g. arsenic or lithium), which acts as an electron donor that creates excess electrons (negative charge) and these occupy the conductance band. P-type semi-conductors are doped with an electron acceptor (e.g. beryllium), which creates 'positive holes' in the conductance band. Positive holes can move around the crystal lattice like conductance band electrons and therefore create conductivity.

Now, if we put an N-type and a P-type together, some of the excess electrons in the N-type cancel out the positive holes in the P-type creating an intrinsic zone of no charge or 'depleted zone'. Solid state α-spectrometers are designed to maximize the width of the depleted zone by applying an electrical field. An α-particle entering the detector deposits its energy in the depleted zone, which causes an electrical pulse proportional to the energy of the decay. Again, if we can amplify that pulse maintaining proportionality, we have spectroscopic resolution of different energy decays.

Now, of course, the high ionization of the α-particle is against us, and α-spectrometry relies on careful preparation of pure single element sources and their deposition as extremely thin films, otherwise there will be significant self-absorption. Sources are usually electroplated onto shiny metal disks and the rule of thumb for the analyst is, if you can see anything on the disk, the sample isn't sufficiently pure or it is too thick for a successful measurement.

Sample preparation

Purifying the sample, let's say getting pure uranium out of a rock, is a job for the chemist, and isotope geochemists are continually working out better ways to get specific elements out of the samples they want to analyse. There are a few options but mostly it is done by chromatography. You might recall doing chromatography in high school by drawing a dot with a ball-point pen on a laboratory filter paper, putting the paper in acetone, and watching the dot separate out into a variety of shades that represent different components of the ink. In the isotope laboratory the ink blot is the sample, the filter paper is ion exchange resin contained in a glass or plastic tube, and the acetone is usually a dilute acid like nitric or hydrochloric. Let's say we are trying to get uranium out of a rock for α-spectrometry. First, we dissolve finely powdered rock in very strong acid—believe it or not, we can do this mainly because a very strong and hazardous acid, hydrofluoric, is able to break up the silicon-to-oxygen bonds that hold rocks together. Once we have it in solution, we pour it onto the ion exchange resin column. Previously, we would have calibrated exactly how much acid we need to add before different elements leave the column. By devising a scheme whereby uranium (and only uranium) comes off the column at a particular point we can get a pure uranium sample in acid solution which we can then partially dry on a hot-plate and electroplate for α-spectrometry.

There is a hitch though; it is reasonably straightforward to get pure uranium but you usually lose a bit and the amount you lose tends to vary from sample to sample, column to column, and day to day. So you need a way of knowing how much you've lost, and the best way to do this is to add an isotope of uranium (or whatever other element you are analysing) that wouldn't be in the original sample—a so-called spike or tracer. Typically, the spike isotope will be a radioactive isotope with a sufficiently short half-life that it doesn't occur in natural samples but was recently created in a nuclear reactor. In the purification, you might lose uranium but you will lose the same amount of spike uranium as sample uranium. So, if you carefully measured how much spike you added by weighing it and you know the uranium concentration of the spike solution, you can measure the concentration of the spike isotope and calculate the concentrations of the natural uranium isotopes in your sample. Examples of spikes used for α-spectrometry include ^{232}U to measure ^{234}U, ^{235}U, and ^{238}U; ^{208}Po to measure ^{210}Po—actually used to calculate ^{210}Pb by assuming radioactive equilibrium; and ^{236}Pu or ^{242}Pu to measure ^{238}Pu, ^{239}Pu, and ^{240}Pu.

We will return to measuring isotopes with mass spectrometry in Chapter 6, but meanwhile let's take a look at the way isotopes are used in medical and biological applications.

Chapter 5
Physics heal thyself: isotopes in medicine

Stable and radioactive isotopes are used extensively in both diagnostic and therapeutic medical applications. Such applications include studies of human body composition, energy balance, protein turnover, and metabolism. Ionizing radiation is key to a host of medical imaging techniques and radioactive isotopes are widely used to target and kill cancer cells. Indeed, few branches of medicine are not impacted in some way by isotopes.

Isotope enrichment

Stable isotope applications of isotopes usually depend on using 'enriched' isotopes as biological tracers. By 'enriched' we mean that the isotopic composition of an element has been manipulated to be different from that of the same element in its natural occurrence. For example, zinc is a common essential mineral in the body which is necessary for the correct operation of the immune system. Zinc also plays a role in cell growth, wound healing, and the breakdown of carbohydrate. Zinc is also necessary for the senses of smell and taste to function correctly. So, imagine we have a patient complaining of reduced or altered flavour of food (flavour is a combination of taste and smell). A doctor might decide to investigate how the patient's body was metabolizing the zinc intake from her diet. 'Normal' zinc in the environment has the following isotopic make-up—^{64}Zn, 49 per cent; ^{66}Zn, 28 per cent; ^{67}Zn, 4 per cent; ^{68}Zn, 19 per cent.

Medical isotope suppliers sell zinc with ^{67}Zn enriched to 80–90 per cent, so very different from natural zinc. This zinc tracer would be introduced into the patient by ingestion in some form of drink or food, or injection into the bloodstream. Subsequent monitoring of the isotopic composition of Zn in blood and urine would show how the element is behaving in the body. Dilution of the tracer signal by natural composition zinc would indicate the distribution of zinc in the body; and the rate of excretion of the tracer Zn would indicate how Zn is being absorbed by the body. As a simple example, if the tracer Zn emerged from the body immediately, this would indicate that zinc absorption was suppressed and the patient's symptoms might indeed indicate a failure to process zinc properly.

We have used zinc to explain the principles behind enriched stable isotope labelling, but other elements, particularly magnesium, iron, and calcium, have been used extensively to study mineral metabolism. However, despite an extensive amount of academic literature, the way the body processes these essential elements remains only partially understood. One of the obstacles to research is the high cost of enriched tracers in the quantities needed in whole-body studies. A response has been to utilize the ability of accelerator mass spectrometry (AMS) to detect tiny amounts of rare isotopes. So, in studies of calcium isotopes aimed at understanding osteoporosis, rather than use large quantities of ^{42}Ca and ^{44}Ca, some studies have used the ^{41}Ca isotope, which is naturally very rare but readily detected by AMS at low levels of enrichment. Similarly, pharmaceutical testing that previously used large amounts of enriched ^{13}C, now typically uses low levels of ^{14}C (well below what would be a radiological hazard) and AMS detection. Indeed, outside purely academic research, the main operators of AMS are in the pharmaceutical industry.

Doubly labelled water

An imaginative variant of isotopic enrichment is the 'doubly labelled water' method. This depends on the different metabolic behaviour of hydrogen and oxygen, which enables us to calculate the rate of metabolism. The subject (the method is widely used in other animals as well as humans) is dosed with ^2D and ^{18}O—in humans this would simply be a drink of heavy water ($^2D_2^{18}O$)—the test then monitors the isotopic composition of saliva, urine, or blood to measure the elimination rate of the tracer.

Cellular respiration breaks down carbon-containing molecules such as sugars to release energy and carbon dioxide (CO_2) which is breathed out. However, in general, food molecules do not contain two oxygen atoms for each carbon atom. So, extra oxygen is needed and this comes from body water. This means that the ^2D introduced as heavy water is excreted as urine but the ^{18}O component of heavy water leaves the body as both excreted urine *and* exhaled breath. From the ^2D we can calculate the amount of ^{18}O lost as body water. The 'missing' ^{18}O is in the CO_2 lost through respiration and, since respiration is the only mechanism to lose CO_2, this, in turn, reveals the rate of metabolism. Typically, the doubly labelled water method is used to establish the average (or field) metabolic rate over days or weeks after administration of the heavy water. As well as human studies, over 200 species of animals have been studied for metabolic rate using this method. Critically, the method is non-radioactive and non-toxic (at the levels administered), so it is suitable for application in humans (including children) and animals without ethical concern.

One of the key conditions in which the doubly labelled water method can be used is in the diagnosis of type 2 diabetes which has become a major health issue in the developed world. In the UK, type 2 diabetes is set to double by 2035 on current trends and already costs the National Health Service £1.5 million per hour—that's roughly the average UK yearly salary per minute! In type 2 diabetes, cells develop a reduced sensitivity to insulin, with an attendant decrease in metabolic rate that can be detected with doubly labelled water.

^{13}C-labelled urea

Another stable isotope test has revolutionized the diagnosis of predilection to stomach and duodenal ulcers. In the early 1980s, a connection was made between ulceration and the presence in the gut of the bacterium *Helicobacter pylori*. Thus, the ability to detect *H. pylori* allows early intervention to prevent ulcers. *H. pylori* has an unusual respiration in which the enzyme urease is used to convert urea to ammonia and CO_2. A ^{13}C breath test for *H. pylori* is based on this phenomenon. A background level of ^{13}C is established in the breath of the subject who is then fed ^{13}C-labelled urea. The breath is tested again ten to thirty minutes later. If the level of ^{13}C has increased, the urea has been broken down and the presence of *H. pylori* is confirmed. The real advantage of this test is that it is completely non-invasive, making it ethically suitable for use on children. In our more deprived communities *H. pylori* is established in the guts of children by their early teens, and the breath test allows medical intervention at the earliest possible opportunity.

Radiotherapy

The medical uses of ionizing radiation are manifold, including X-ray imaging and radiotherapy with external X-ray beams. Such applications tend to generate X-rays from linear accelerator instruments rather than discrete isotopic sources. However, there is a substantial branch of radiotherapy that does utilize radioactive isotopes, particularly in the treatment of cancer. The methods work because radiation damages the DNA of cells. But, of course, both cancerous and healthy cells are susceptible to radiation damage, so, in these treatments, a key priority is to maximize delivery of the radioactive source to the tumour cells while minimizing the effect on adjacent healthy cells. Let's look at some of the imaginative ways that radiation oncology, the field of medicine dedicated to treating cancer with radiation, has devised for delivering radiation to tumours.

Brachytherapy

Perhaps the most straightforward method of localizing the delivery of radiation is simply to place radioactive sources beside the cancer cells to be destroyed—a variety of procedures collectively known as brachytherapy or internal radiotherapy. This can be carried out as an intraoperative procedure associated with surgical removal of a tumour and designed to prevent recurrence of the cancer. However, there are brachytherapy methods that are less surgically invasive, particularly in the treatment of vaginal, uterine, and cervical cancer (in women) and prostate cancer (in men). In radiotherapy for cervical cancer, small radioactive metal balls are introduced to the cervix using applicator devices similar to a long thin plastic tampon. The applicators are inserted into the cervix and a magnetic resonance imaging (MRI) scan or computerized tomography (CT) scan is used to ensure correct positioning of the applicators. The patient is then connected to a brachytherapy machine which delivers the radioactive sources to the applicators located to deliver radiation to the tumour with minimal effect on neighbouring healthy cells. Internal radiotherapy is administered in two distinct therapeutic strategies: low dose-rate (LDR) brachytherapy delivers a relatively low radiative dose but the sources, typically ^{137}Cs (caesium) in cervical cancer treatment, remain in place for a few days. High dose-rate (HDR) treatment involves a more active source, typically ^{60}Co (cobalt) or ^{192}Ir (iridium), which remains in place for only a few minutes per session but is usually repeated over a few days.

Brachytherapy is also widely used to treat cancer of the prostate gland. Here, the radioactivity is delivered on 'seeds' about the size of a grain of rice which can also be joined together to make radioactive strands. The seeds are delivered through needles inserted into the perineum (the area between the scrotum and rectum). Guided by an ultrasound scanner the needles are located using an external grid device that ensures the radioactive seeds are positioned according to a pre-determined 'map reference' corresponding to the shape of the tumour. In any one treatment about 70–150 radioactive seeds are introduced. The radioactivity emitted by the seeds has a very short penetration path ensuring that radioactivity is limited to the target tumour, with minimal effect on adjacent tissues. Again, prostate cancer brachytherapy can be LDR, typically using ^{125}I (iodine) or ^{103}Pa (palladium), or HDR, using ^{60}Co, ^{192}Ir, or ^{169}Yb (ytterbium). LDR sources are inserted to the prostate permanently but have short half-lives and decay away over a few weeks or months. HDR brachytherapy involves administration of the radioactive seeds for only a few minutes, with treatment repeated two or three times over a period of a few days.

Targeted radionuclide therapy

A variant of internal radiotherapy uses the body's natural blood circulation to deliver a localized dose of radioactivity. The liver is unique in that it receives oxygenated blood from the hepatic artery and partially de-oxygenated blood from the hepatic portal vein in the ratio of about one to three. By contrast, some cancers of the liver receive all their blood from the arterial route. Thus, it is possible to deliver radiotherapy preferentially to the tumour via the hepatic artery using millions of tiny beads of either glass or plastic coated in a radioactive isotope. ^{90}Y (yttrium) which undergoes a 2.28 MeV (mega electron volt) β decay to ^{90}Zr with a sixty-four-hour half-life is commonly used as the radioactive source in treatment of liver cancer. ^{90}Y bead treatment is usually palliative—aimed not at curing hepatic tumours but at enhancing quality of life and prolonging survival by a few months.

Not all liver cancers are amenable to ^{90}Y treatment, with the main consideration being how specifically the radioactivity can be delivered to the tumour. Prior to radiotherapy it is necessary to determine the nature of blood flow to the liver and other organs (e.g. lungs). A so-called contrast dye that is easily traced by X-ray is introduced by a catheter through the groin into the hepatic artery. X-ray images then track the movement of the dye—a process known as a 'mapping angiogram'. If the blood flow stays predominantly within the liver without substantial loss to other organs, the tumour can be treated with ^{90}Y beads. Suitable patients have ^{90}Y delivered to the liver by injection into the hepatic artery and localization of the radioactivity can be enhanced by blocking some blood vessels and using others specifically to deliver the ^{90}Y microspheres to the tumour.

Another approach to delivering radiotherapy selectively to tumours is to choose an isotope of an element that has a particular affinity for the organ in which the tumour is located. Iodine has a strong affinity for the thyroid gland to the extent that in the aftermath of nuclear weapons attacks or nuclear accidents it is best practice to dose survivors with non-radioactive natural ^{127}I to minimize thyroid uptake of radioactive ^{131}I and ^{129}I, which are fission product isotopes. Following the same rationale, ^{131}I is used to deliver radioactivity to the thyroid to treat cancer in that organ. On the face of it there is a fundamental paradox here as we are proposing to deliver to the diseased thyroid the very isotope that is implicated in causing thyroid cancer. However, the treatment depends on the fact that the cancer cells are replicating at a much faster rate than normal cells. A single dose of iodine will not only target the thyroid but will also be preferentially toxic to proliferating cells of the tumour.

As we saw in Chapter 3, strontium (Sr) has a strong tendency to mimic calcium, thus radioactive ^{89}Sr can be used to treat cancer patients with painful bone metastases. Delivered by infusion to the bloodstream or simply ingested, ^{89}Sr delivers β-particles targeted at the malignant regions where the turnover of calcium is highest. Radium is also an analogue element of calcium, being another of the alkali earth metals or group 2 elements of the periodic table (Figure 2). ^{223}Ra treatment is used particularly to treat prostate cancers that have spread to the bones. Introduced into a vein through a cannula, the ^{223}Ra concentrates in the bones selectively at cancerous sites. ^{223}Ra is an α-particle emitter with an 11.4-day half-life, so it delivers high energy radiation, but each α-particle penetrates only a few cells making it a highly targeted radiotherapy.

Rather than rely on the natural tendency of a radioactive isotope to associate with a particular tissue, another therapeutic strategy is to link the radioisotope to a molecule that guides the radioactivity to the target tissue. Such approaches are at the forefront of cancer research developing 'smart' drugs that can target and attach to cancer cells providing radiotherapy that exclusively attacks cancer cells. ^{177}Lutetium DOTA-TATE (LuDO) is a somatostatin analogue drug carrying radioactive ^{177}Lu. The cell surfaces of neuroendocrine tumours (i.e. tumours of the nervous and hormonal systems) express somatostatin receptors. LuDO seeks out these receptors, it binds to the tumour, and the radioactive lutetium destroys the rogue cells.

The drug Ibritumomab tiuxetan (known commercially as Zevalin) uses tiuxetan as a chelating agent to join ^{90}Y to monoclonal mouse IgG1 antibody (Ibritumomab). This antibody binds to the CD20 antigen found on both normal and malignant B cells. B cells are lymphocytes in the immune system that are involved in the creation of antibodies. The human body produces millions of B cells daily and these circulate in the blood and lymph performing an immune surveillance function. B cells are susceptible to malignant transformation. Zevalin seeks out B cells and delivers the ^{90}Y β-particle killing both normal and malignant cells, and allowing repopulation by new non-cancerous B cells.

Metaiodobenzylguanide labelled with ^{131}I (^{131}I-MIBG) treats neuroblastoma, which is a childhood cancer that frequently originates in the adrenal glands. ^{131}I-MIBG, a molecule that is similar in structure to noradrenaline, targets adrenergic tissues and delivers the ^{131}I β-decay with an eight-day half-life causing cell death. A small amount of deiodination of ^{131}I-MIBG occurs in the circulation. Of course, the radioiodine also has a strong affinity for the thyroid gland so, much as in the aftermath of nuclear accidents, it is common practice to deliver ^{131}I-MIBG along with a thyroid blocker such as non-radioactive ^{127}I as potassium iodide, to overwhelm the thyroid to avoid uptake of radioactive iodine.

Imaging

A variant of MIBG treatment is used to image neuroendocrine tumours. In this case the radioisotope attached to MIBG is ^{123}I which is a 13.22-hour half-life γ-ray emitter. The γ-rays do little cell damage but travel through the body and can be detected with a gamma camera. Again the iodine is concentrated in adrenergic tissue and creates an image of the neuroendocrine tumour. MIBG imaging is one of a range of imaging techniques grouped as Single Photon Emission Computed Tomography (SPECT). Rather than producing images using an external source of X-rays, SPECT images result from γ-rays actually produced within the body. SPECT produces a series of two-dimensional (2-D) images of the three-dimensional (3-D) distribution of γ-activity but by taking multiple images at different angles the computer tomography builds up a true 3-D representation.

Various molecules labelled with a range of radioactive isotopes are used to attach to different organs and produce images that yield a variety of information. The metastable γ-emitting technetium isotope ^{99m}Tc is frequently used bound to molecules that target the tissue of interest. For example, ^{99m}Tc-HMPAO (hexamethylpropylene amine oxime) is taken up by brain tissue in a manner that reflects the flow of blood in the brain. This, in turn, indicates local metabolism that can be used to diagnose conditions such as dementia and Alzheimer's disease. SPECT is able to differentiate between the localized brain damage resulting from several stroke episodes compared to the more continuous damage associated with Alzheimer's.

Another tomographic imaging method that utilizes a radioactive isotope is positron emission tomography (PET). This method uses fluorodeoxyglucose (FDG) as an analogue of glucose labelled with the radioactive isotope fluorine-18. The uptake of glucose indicates high tissue metabolic activity such as that associated with tumour growth. In FDG the oxygen atom that is replaced by ^{18}F is required in all cells to metabolize glucose, so further metabolism of glucose is prevented. Hence, once a cell is labelled with ^{18}F, it stays labelled until the ^{18}F decays through radioactivity (half-life is 109.8 minutes).

^{18}F decays dominantly by positron emission. Each positron travels less than a millimetre before losing enough energy to interact with an electron. When the positron meets an electron, both particles are destroyed generating a pair of annihilation gamma photons which travel in opposite directions to be detected by a material called a scintillator that generates a photon of light when impacted by the γ-rays. The photons are collected and multiplied by a photomultiplier tube (see Chapter 4) similar to those used in image-intensifying camera lenses used by wildlife photographers to film at night. The technique recognizes photons attributable to ^{18}F decay when they arrive in opposite directions simultaneously (i.e. within a few nanoseconds). Background photons that don't arrive in coincidence are ignored. The resultant ^{18}F image tomographically combines a series of 2-D slices to form a 3-D map of tissue metabolic activity. Modern scanning instruments now often combine PET with X-ray CT or MRI scanning to provide both anatomic and metabolic information.

Radiotherapy in the future

Radiation therapy has become an important tool in the armoury of the oncologist or cancer specialist. Today, there is an ongoing effort to design new treatments that will improve cancer survival and the quality of life of cancer patients. In particular, efforts continue to optimize dose fractionation in which the radioactive isotopes distinguish and seek out tumours but avoid healthy cells. So-called 'conformal' radiation techniques aim to provide 3-D delivery of radioactivity designed to match the actual shape of individual tumours.

Increasingly, radiotherapy is optimized in combination with other cancer treatments. In particular, recent years have seen the emergence of mechanistic biological studies in which quantitative and computational methods, inherited from the physical sciences, are used to explore the interactions that contribute to a full understanding of the molecular mechanisms that underpin life. Undoubtedly, radiotherapy will continue to play a major role in cancer treatment as we increasingly understand cell death pathways, optimize the interaction with other treatments, and maximize dose fractionation.

Chapter 6
Measuring isotopes: mass spectrometers

Nuclear spectroscopy, the broad range of techniques that were discussed in Chapter 4, has two main limitations. First, it can only be used to detect radioactive isotopes, so it isn't able to provide any information about the distribution of stable isotopes like ^1H, ^2H, ^{12}C, and ^{13}C. Second, the techniques detect the decay of a radioactive isotope, so for isotopes with longer half-lives, long count times are required to achieve precise measurements. For isotopes with half-lives that allow us to measure geological timescales (i.e. millions to billions of years) the necessary count times are totally impractical. We need a technique that can count the atoms of different isotopes without relying on their radioactive decay, and that technique is mass spectrometry.

The first mass spectrometer was designed by J.J. Thompson and his research assistant F.W. Aston. In 1912, Thompson and Aston channelled a stream of neon ions through magnetic and electric fields and recorded their deflection by exposing a photographic plate. The resultant image showed two patches of exposure corresponding to neon. They had discovered that neon has two stable isotopes (although of course, Soddy was yet to coin the term), ^{20}Ne and ^{22}Ne, and demonstrated that, as well as radioactive isotopes, there were stable atoms of the same element that had different mass.

Mass spectrometers have become routine laboratory instruments in a variety of disciplines and the famous American scientist, Philip Abelson, once wrote that the mass spectrometer was 'probably the most ubiquitous tool in the Earth Sciences'. Here we will concentrate on mass spectrometers as tools to quantify the abundance of different isotopes. However, since we know the natural abundance of many stable isotopes, a mass spectrometer can be used to quantify the concentration of a particular element by monitoring an isotope of the element of interest that is not overlapped by isotopes of other elements.

Isobars

For example, let's take iron (Fe) which has four stable isotopes ^{54}Fe, ^{56}Fe, ^{57}Fe, and ^{58}Fe—actually ^{54}Fe is radioactive but with a half-life of over 10^{22} years it is 'observationally stable'. Also at mass 54 is an isotope of chromium (Cr). ^{54}Fe and ^{54}Cr have the same cardinal masses and are referred to as isobars. In fact, ^{54}Fe and ^{54}Cr are slightly different in mass (53.9383 and 53.9389 atomic mass units, respectively) but a mass spectrometer capable of making precise isotope ratio measurements would not be able to resolve such small mass differences. The mass 54 peak would be a mixture of Cr and Fe, and ^{54}Cr is an 'isobaric' interference on the ^{54}Fe peak. To measure Fe alone, you would avoid mass 54. Similarly, ^{58}Fe is interfered by the most abundant isotope of nickel (^{58}Ni), so we would avoid mass 58. No elements other than iron have stable isotopes at masses 56 and 57. Hence, both would be suitable analytical targets to quantify total iron content. In practice, ^{56}Fe would be the isotope of preference because it constitutes about 92 per cent of total iron whereas ^{57}Fe is only 2 per cent, so the 56 signal would be forty-six times that from 57. If we quantify ^{56}Fe and we know that 92 per cent of iron is ^{56}Fe, then total iron will be ^{56}Fe/0.92.

Mass spectrometer—one name, many variants

Returning to basics, all mass spectrometers have three essential components—an ion source, a mass filter, and some sort of detector (Figure 12). Let's restrict ourselves to the sub-classes of mass spectrometer that are used to measure isotope ratios and discuss the different approaches to creating these basic components.

Gas source isotope ratio mass spectrometers

The light stable isotopes (H, C, N, O, S) are all readily introduced into the mass spectrometer as gases—usually H_2, CO_2 (for C and O), N_2 or N_2O (for N and O), and SO_2 or SF_6 (for S). The first requirement is that the gases need to be purified, and the processes used vary dependent on the nature of the samples to be analysed and the element of interest. For example, calcium carbonate readily dissociates when exposed to even weak acids, liberating CO_2. So, a standard methodology for analysis of $\delta^{13}C$ and $\delta^{18}O$ in calcium carbonate shells simply dips the sample into a bath of phosphoric acid, collects the CO_2 evolved, and injects it into the gas source mass spectrometer. By contrast, silicate rocks

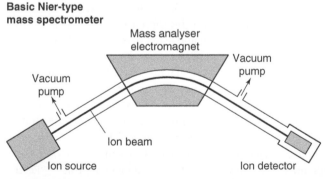

Basic Nier-type mass spectrometer

Mass analyser electromagnet

Vacuum pump

Vacuum pump

Ion beam

Ion source

Ion detector

12. **The essential elements of a mass spectrometer.**

like granite and basalt or minerals like quartz and feldspar react very slowly and incompletely with even the strongest mineral acids. Instead, oxygen can be liberated from silicates by reaction with very aggressive halogen compounds like chlorine trifluoride (ClF_3) or bromine pentafluoride (BrF_5). Both these reagents are highly explosive in air, so evacuated preparation lines for O-isotope analysis of silicates need to be rigorously designed, and carefully risk-assessed and maintained. Today, the preferred method is carefully to heat a few silicate crystals with a laser in the presence of minimal quantities of ClF_3.

In a gas source isotope ratio mass spectrometer ionization is achieved by passing a current through a wire filament. Much in the way that a light bulb emits photons, thermionic emission generates electrons which are then concentrated into a beam by being attracted towards an electrode. The electron beam is directed towards the sample gas and accelerated to an energy that maximizes the ionization efficiency of this 'electron impact' ionization process.

Thermal ionization mass spectrometers

Heavier elements such as the alkali metals, alkali earths, and rare earths are typically ionized by thermal ionization mass spectrometry (TIMS). To minimize isobaric interferences and to maximize ionization efficiency, TIMS sources need to be highly purified. As described in Chapter 4 for α-spectrometry, purification is usually achieved by ion exchange chromatography. Processing of the sample is performed in an ultra-clean laboratory usually supplied with filtered air that creates an over-pressured cleanroom designed to prevent dirty air entering the lab. Chemical reagents are ultra-clean, and most laboratories take commercially available acids and clean them further by sub-boiling distillation in Teflon stills. Laboratory water is totally deionized by reverse osmosis systems costing thousands of pounds, and the lab scientists typically wear cleanroom overalls, hoods, overshoes, and gloves, which serve the dual role of preventing sample contamination and protecting the chemist from hazardous strong acids—the most challenging samples need the most hazardous mineral acids (e.g. hydrofluoric and perchloric to effect dissolution).

All these provisions are necessary because we are dealing with tiny amounts of sample material and trying to extract elements which are also present in the environment without contaminating the sample. For instance, in my lab we analyse lead (Pb) in rocks like basalt. Concentrations are typically a few parts per million (ppm) or microgrammes (µg) per gramme (g), and we digest about 50 milligrammes (mg) of powdered rock. So, for a 5 ppm rock, assuming we extract 100 per cent of the Pb, the final sample is $50 \times 10^{-3}g \times 5 \times 10^{-6}g$ (i.e. $2.5 \times 10^{-7}g$ or 250 nanogrammes (ng)). I like to keep the ratio of sample to 'blank' (i.e. contamination) greater than 1,000/1, so no more than 0.1 per cent of the final samples is environmental contamination, so we need blanks that are less than 250 picogrammes (pg). More challenging applications require even better blanks and it is possible to push Pb to a few picogrammes. Actually, Pb blanks have become easier since the abolition of lead additives in petrol. However, my lab is about 50 metres (m) down the corridor from the nuclear spectroscopy lab where there are industrial amounts of Pb shielding, so I need to be careful where I go and what I touch!

TIMS samples emerge from the chemistry laboratory as a tiny spot of the element of interest usually as a chloride or nitrate in the bottom of a Teflon beaker. We pick the sample up in a microlitre of acid and place it on a mass spectrometer filament made of a metal capable of being heated to high temperature without melting and with a relatively low tendency for thermionic emission of electrons, typically rhenium, tantalum, or tungsten (Figure 13). We dry the sample by putting a current through the filament and then load the sample into a barrel

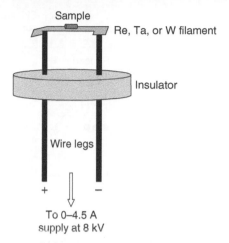

Sample

Re, Ta, or W filament

Insulator

Wire legs

+   −

To 0–4.5 A
supply at 8 kV

13. A filament assembly used in a thermal ionization mass spectrometer (TIMS).

which carries as many as twenty samples that can be analysed one after another. Loading several samples at once reduces the number of times the mass spectrometer ion source is vented to atmospheric pressure between barrels. The size of individual barrels is largely dictated by the volume of the ion source that can be pumped to a working vacuum in a convenient length of time.

TIMS ion sources are limited because only a handful of elements will ionize as they evaporate off a hot filament. One approach for 'difficult' elements is to load them onto a filament held at relatively low temperature from which neutral atoms evaporate, and ionize them with a second filament held at higher temperature. Similarly, there are a few chemical methods designed to increase the ionization efficiency of TIMS ion sources but the method is really restricted to elements with relatively low ionization energies (e.g. Sr, Nd, Pb).

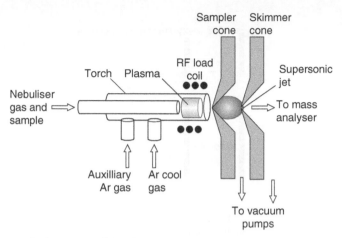

14. The ion source of an inductively coupled plasma mass spectrometer (ICP-MS).

Inductively coupled plasma mass spectrometers

By contrast, inductively coupled plasma (ICP) ion sources are able to ionize the majority of elements. The ICP ion source works by electrically inducing a plasma in a stream of argon gas flowing through a quartz torch (Figure 14). The sample, either dissolved in acid or in a stream of inert gas, is introduced to the plasma where it is ionized at a temperature of about 6,000°C and the ions deflected by electrostatically charged lenses into the mass spectrometer. The big advantage of the ICP source is that it is able to ionize efficiently all the elements whose ionization energy, or the amount of energy required to remove the outermost electron to form an ion, is lower than that of the argon gas forming the plasma. Theoretically, only H, He, F, and Ne will not ionize in the argon plasma. However, the big disadvantage of the ICP sources is also how readily it forms ions. Quite a lot of elements have second ionization potentials lower than the first ionization potential of argon (i.e. they will form 2+ ions in the ICP source). All mass spectrometers are actually mass (m) over charge (z) ratio spectrometers. So, 2+ ^{88}Sr will occupy the same m/z space as 1+ ^{44}Ca. Equally, the energy of the ICP source is high enough to ionize molecules, so ^{44}Ca$_2$ would interfere with Sr at mass 88.

Actually, the ICP ion source has such high energy that some very surprising potential isobaric interferences are generated. Returning to the example of iron we used before, in addition to the elemental isobars from Cr and Ni, ICP sources generate molecular interferences such as ArN (mass 54 and 55) and ArO (56, 57, 58). Instrument manufacturers have come up with various ways to overcome these interference effects but, in short, the ICP-MS analyst needs to be constantly conscious that cardinal mass peaks may be made up of more than simply the elemental isotope of that mass.

Most ICP-MS instruments are used for elemental analysis by carefully choosing isotopic peaks that aren't affected by isobaric interferences. However, with appropriate ion optics, ICP sources are used for isotope ratio measurements of elements whose high ionization energies make them inaccessible by TIMS. Furthermore, ICP sources can be supplied with sample material using a laser beam fired at sample materials loaded in a chamber through which gas is flowed to transport the sample to the ICP source. Such *in situ* laser-ablation measurements can be made with a spatial resolution of a few tens of microns (10^{-6}m). In geology, such capability is used to examine isotopic variability within individual mineral crystals.

Secondary ion mass spectrometers—ion microprobes

Secondary ion mass spectrometry (SIMS) ion sources use a beam of charged ions to 'sputter' ions from the surface of a target. For example, negative ions of oxygen can be fired at materials mounted as polished targets from which positive ions are extracted and injected into a mass spectrometer. The oxygen ion beam can be focused to spot sizes of 10–30 microns and again are used to look for isotope variations with high spatial resolution. The (S)ensitive (H)igh (R)esolution (I)on (M)icroprobe (SHRIMP) pioneered by Prof Bill Compston at the Australian National University is famous for U-Pb dating individual crystals of the mineral zircon. Similarly, SIMS instruments are becoming used to probe the isotopic composition of individual cells in biological systems.

Of course, there is a play-off for high spatial resolution because very small amounts of sample material are analysed. Just as we saw for nuclear spectroscopy, even a perfect mass spectrometer is limited in precision by the number of atoms counted. So, U-Pb dates on whole crystals of zircon are more precise than spatially resolved SIMS measurements of tiny zones within the crystals. However, the SIMS measurements can reveal internal age-structure that is invisible to TIMS. Typically, even individual zircon grains show age zonation that shows multiple episodes of crystal growth that can be separated by hundreds of millions of years and preserved for billions of years—truly, zircons are forever.

Ion optics

The most ubiquitous mass analyser for isotope ratio mass spectrometry is a simple electromagnet. Ions from the ion source are electrically accelerated into the flight tube of the mass spectrometer through a series of electrostatic lenses that focus ions much in the same way that spectacles focus light into the wearer's eye (Figure 15). The ion beam forms an image of the final focus slit which is focused on the detector. Individual ions have kinetic energy which is dependent on the voltage used to accelerate the ions and their mass. So, ^2H ions have twice the energy of ^1H ions. When the ions enter the magnetic field between the poles of the electromagnet they are deflected but the heavier isotopes are affected less than the lighter isotopes. This separates the different isotopes generating a mass spectrum. Mass spectrometers with a

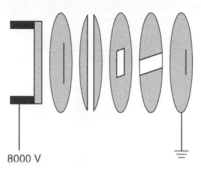

8000 V

15. The filament and lens stack from a TIMS—filament is held at 8,000V and image slit is at earth potential. The intermediate lenses can be varied in voltage in order to focus the ion beam on the detector at the other end of the mass spectrometer.

single detector vary the magnetic field of the electromagnet to collect each isotope ion beam one after another. However, these days many mass spectrometers have multiple detectors, so that each isotope has a dedicated detector. So long as the individual detectors can be cross-calibrated better than time-dependent fluctuations in the rate of ion production, a multiple detector instrument will give improved measurement precision and achieve a given precision in a much shorter period of time.

Detectors

There are two main types of detector used to quantify the relative abundance of different isotopes. Large signals are measured using a Faraday detector (Figure 16). This is a small cuboid of conducting material with dimensions of about 5 mm (width) × 15 mm (height) × 20 mm (depth). The Faraday is connected to a very robust earth potential. This ensures that every ion impacting the Faraday is neutralized by an electron from the earth. The currents involved are tiny and defy measurement. For example, a million ions per second generate a current of 1.6×10^{-13} A (amps) which can't be measured as an analogue current. So, the stream of

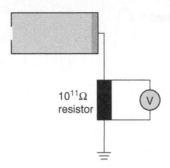

16. Schematic of a Faraday detector—ion current is quantified by measuring the voltage across a high ohmic resistor connected to earth potential generated by electrons flowing from earth to neutralize the ions impinging on the Faraday cup.

$10^{11}\Omega$
resistor

V

electrons from earth is channelled through a high gain amplifier. Typically the feedback resistor in the amplifier circuit will have a resistance of 10^9 to $10^{12}\Omega$ (ohms). Ohm's law states that voltage (V) is proportional to current (I) multiplied by resistance (R) i.e. V=I.R. So, for a $10^{11}\Omega$ resistor, 1.6×10^{-13}A generates $1.6 \times 10^{-13} \times 10^{11}$ volts (i.e. 16 mV), which is readily measured.

Small isotope signals are measured using a variant of the photomultiplier tubes (PMT) we already encountered in nuclear spectroscopy but instead of responding to a photon, the first dynode generates a few electrons in response to being impacted by an ion. The resultant cascade of electrons generates a measureable pulse in response to each ion arrival. Small signals, down to individual ion arrivals can be measured by taking the multiplier output through the same digital pre-amplifiers, pulse-height discriminators, and counters used in nuclear spectroscopy.

Abundance sensitivity

It is often said that 'nature abhors a vacuum' and this is a challenge to the mass spectrometrist. Everything we have described about mass spectrometry has to take place in vacuum—typically 10^{-8} to 10^{-10} millibars. To put this in context this is about a billion times lower vacuum than a domestic vacuum cleaner. We use a variety of pump devices to achieve high- and ultra-high-vacuum and mass spectrometers are regularly baked to 300–400°C to purge gas adsorbed onto the metal surfaces of the instrument flight tube. Mass spectrometers need to achieve high vacuum to allow the uninterrupted transmission of ions through the instrument. However, even high-vacuum systems contain residual gas molecules which can impede the passage of ions.

Even at very high vacuum there will still be residual gas molecules in the vacuum system that present potential obstacles to the ion beam. Ions that collide with residual gas molecules lose energy and will appear at the detector at slightly lower mass than expected. This tailing to lower mass is minimized by improving the vacuum as much as possible, but it cannot be avoided entirely. The ability to resolve a small isotope peak adjacent to a large peak is called 'abundance sensitivity'. A single magnetic sector TIMS has abundance sensitivity of about 1 ppm per mass unit at uranium masses. So, at mass 234, 1 ion in 1,000,000 will actually be ^{235}U not ^{234}U, and this will limit our ability to quantify the rare ^{234}U isotope.

Accelerator mass spectrometers

As we have already discovered, only 1 carbon atom in 10^{12} is ^{14}C. In a conventional isotope ratio mass spectrometer the high mass tail from ^{13}C swamps any signal from ^{14}C. Until about twenty years ago our inability to resolve ^{14}C by mass meant that radiocarbon dates were produced by liquid scintillation counting of the ^{14}C β-decay to ^{14}N. However, the half-life of ^{14}C is over 5,000 years, so counting decays is pretty slow work. What if we could design a mass spectrometer that can resolve one atom of ^{14}C from 10^{10} atoms of ^{13}C? Enter accelerator mass spectrometry (AMS).

AMS instruments use very high voltages to achieve high abundance sensitivity. Carbon samples are introduced to the AMS as graphite targets pressed into a tiny hole in a metal cathode. The ion source is similar to that used in the ion microprobe but instead of using O_2^- to sputter positive ions, the AMS uses a source of positive caesium ions to sputter negative carbon ions from the graphite target. Serendipitously, nitrogen (the main mass 14 isobaric interference) does not form stable negative ions, so a major impediment to radiocarbon mass spectrometry is overcome immediately. Initially, singly charged negative carbon ions are accelerated by biasing the ion source up to 60 kV below ground potential. An electromagnet acts as the first mass filter directing ions into the accelerator 'tank'. This is a metal container about 10 m long in which motor driven metal belts generate a charge of up to 5 million volts (MV) to an electrode. The tank contains SF_6 gas which is a strong electrical insulator—air-insulation is ineffective above about 1 MV.

The carbon ion beam is attracted to the 5 MV 'stripper canal' electrode whereupon electrons are stripped from the negative ion to produced positively charged ions, typically C^{4+}. Removing electrons has the secondary effect of dissociating any molecules. So, a molecule like $^{13}C^{1}H$ which would be an isobaric interference for ^{14}C is reduced to separate carbon and hydrogen atoms. Now, our carbon ion has 4+ charge and is repelled from the 5 MV electrode with 20 MeV of energy.

As 20 MeV ions move really quickly we need a big (5 tonne) magnet to bend them and be the mass filter that separates ^{14}C, ^{13}C, and ^{12}C. Now we need to remember that 'mass spectrometers' are actually mass to charge ratio filters because, even in the case of this 40 m long instrument, they can't separate $^{14}C^{4+}$ from $^{7}Li^{2+}$ or $^{7}Li_2^{4+}$. However, carbon and lithium behave very differently when they enter a gas ionization detector filled with propane. The rate at which different elements lose energy in such a detector is proportional to their atomic number and this property provides the final tool with which to distinguish isobaric elements of different elements.

As I write this chapter, the human population of the world has recently exceeded seven billion. I've already mentioned that one carbon atom in 10^{12} is mass 14. So, detecting ^{14}C is far more difficult than identifying a single person on Earth, and somewhat comparable to identifying an individual leaf in the Amazon rain forest. Such is the power of isotope ratio mass spectrometry.

Chapter 7
Reconstructing the past and weathering the future

Climate records

In Chapter 3 we discussed how the evolution of an air mass from evaporation of ocean water to precipitation of rain and snow is accompanied by progressive depletion in ^{18}O, so, for example, water with $\delta^{18}O$ of 0‰ evaporates water vapour depleted in ^{18}O (e.g. $\delta^{18}O$ = −13‰) and that vapour condenses to form precipitation with higher $\delta^{18}O$ than the vapour and a residual air mass with even lower $\delta^{18}O$. Two factors combine to make $\delta^{18}O$ a sensitive measurement or proxy for former climate change. First, as climate has shifted between glacial and interglacial periods, the amount of water stored on the continents as ice has changed. The ice is depleted in ^{18}O (low $\delta^{18}O$) and therefore ^{18}O is returned to the oceans through (high $\delta^{18}O$) rivers, etc., more efficiently than ^{16}O, forcing the $\delta^{18}O$ of the ocean to increase as temperatures fall and sea level declines during a glacial period.

Thus, if we can reconstruct the $\delta^{18}O$ of the ocean back through time, we should have a robust record of continental ice volumes which can be extrapolated to deduce changes in global climate. Tiny microfossils called Foraminifera form calcium carbonate shells, which, provided they are not affected by *post mortum* exchange, will record the $\delta^{18}O$ composition of the seawater in which they grew. Actually, the Foraminifera shells don't have seawater $\delta^{18}O$, but the process of carbonate formation imparts a small but well-understood O-isotope fractionation so that seawater $\delta^{18}O$ is easily calculated. In fact, it turns out that this second effect, the difference between carbonate $\delta^{18}O$ and seawater $\delta^{18}O$, is itself temperature dependent with lower temperatures increasing the $\delta^{18}O$ of carbonate growing from a given $\delta^{18}O$ water. These two effects work together because both encourage higher $\delta^{18}O$ carbonate to grow under glacial conditions.

Scientists trying to reconstruct past climate collect drill cores of sediments below the sea floor. As one drills deeper into the sea bed, the sediments recovered will be progressively older, having built up over tens of thousands of years. Back in the laboratory, these sediment cores are split into distinct time slices and sieved into discrete grain size fractions. These can then be washed and under a microscope individual Foraminifera are selected for analysis. Because modern mass spectrometers can accurately analyse a few individual microfossils, it is usual to compare samples of the same species of Foraminifera thereby ensuring any inter-species effects are avoided.

In the plot of foraminiferal $\delta^{18}O$ versus age (Figure 17), you will notice that the $\delta^{18}O$ axis scale is reversed so that decreasing $\delta^{18}O$ (and hence increasing temperature) points upwards. On this inverted scale, positive (low $\delta^{18}O$) peaks indicate higher seawater temperatures and hence interglacial periods whereas negative (high $\delta^{18}O$) troughs indicate lower temperature glacial conditions. By combining many such records, a consensus has emerged which geochemists have divided into a series of numbered marine isotope stages. Odd-numbered stages are warm periods, starting with the current interglacial, whereas the even-numbered stages correspond to cold periods. Clearly, the patterns are complex but notice how interglacials descend quickly into glacials over about 10,000 years (or 10 ka) whereas the rise back to interglacial conditions has repeatedly occurred over about 100 ka. Superimposed on this basic pattern are smaller oscillations which clearly

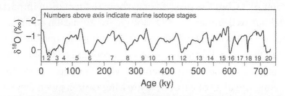

17. A $\delta^{18}O$ record constructed from Foraminifera, similar to those shown in the electron microscope photographs, taken from a deep-sea sediment core.

indicate climate change but are relatively short-lived phenomena. A period of cold temperatures within an interglacial (warm) period is known as a stadial whereas a warm spell within a glacial (cold) period is called an interstadial. We can extract a lot of information from these oxygen isotope records, but let's start with a simple example to introduce the methodology that has been adopted in palaeoclimate studies—here I follow closely the approach described by John Imbrie in his William Smith lecture to the Geological Society of London in 1984.

Gain and phase models

Figure 18 indicates the relationship between the amount of solar radiation energy, so-called insolation, arriving at a northern hemisphere location through the year. Obviously, insolation peaks in the summer and troughs in winter. Also plotted is the sea surface temperature (SST) at the site. The SST follows an identical pattern to insolation but the two are shifted in time by a couple of months; the records are 'out of

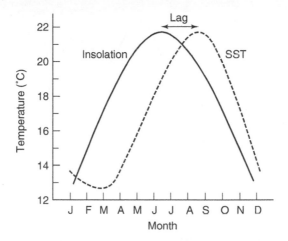

18. A simple northern hemisphere seasonal temperature record showing insolation (solid line) and sea surface temperature (dashed line).

phase'. By preference, I would holiday in the Mediterranean countries in September to avoid the maximum insolation of June and July but take advantage of the same insolation whose effect on SST kicks in and makes September the most comfortable month for swimming in the sea.

Now let's look at a slightly more sophisticated example—areas affected by the Indian monsoon have annual SST similar to that plotted in Figure 19. The annual summer peak in SST still exists but superimposed upon it is a semi-annual cycle which manifests as a temperature minimum in August.

This approach is known as 'gain and phase modelling'. A gain and phase approach to the monsoon situation is illustrated. In this 'variance spectrum' (Figure 20) the statistical variance in SST is presented as a function of the frequency of the variance rather than as a simple time-series (e.g. Figure 19). The area under the

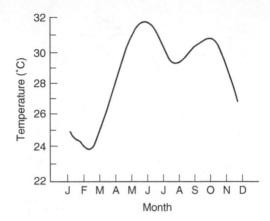

19. Sea surface temperature in a region displaying a strong semi-annual monsoon climate component.

curve represents the total statistical variance in SST with most of the variance (i.e. high variance density) concentrated in two narrow bands representing the annual and semi-annual components of the total response.

Annual and semi-annual cycles are combined to produce a model of SST variation. This is compared to actual measured SST data (in palaeo-records this would be Foraminifera $\delta^{18}O$ variations)

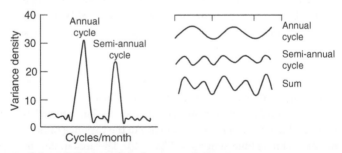

20. Gain and phase model of monsoonal climate.

and the difference between the real data and the model predictions are called the residuals. The extent to which the model fits the data can be expressed as a correlation coefficient or 'r' value. A correlation coefficient of 1 indicates a perfect fit. In our example, r = 0.91 which indicates that 91 per cent of the variation in the data set is explained by the combined model.

Milankovitch orbital cycles

If we apply a similar spectral analysis to a consensus Foraminifera $\delta^{18}O$ record stretching back from the present to almost 800 ka ago, significant peaks in variance density emerge at 100 ka, 41 ka, 23 ka, and 19 ka. So, a large proportion of the $\delta^{18}O$ variation can be explained by a gain and phase model that incorporates these frequencies. Significantly, these are identical to the astronomical periodicities of the Earth's orbit predicted by Milutin Milankovitch.

The 100 ka period relates to the eccentricity parameter which describes the deviation of the Earth's elliptical orbit around the sun from circularity. Through time, the orbit changes from near-circular to more elliptical and, while the exact details are a little complex, the overall effect is a 100 ka periodicity. The degree of eccentricity determines how close the Earth is to the sun during the year. A circular orbit would minimize variations in insolation whereas a large eccentricity will vary insolation through the year.

The Earth's rotational axis is not at right angles to its orbital plane around the sun. In other words, the Earth is tilted in space so that the Earth's equatorial plane is not coplanar with the ecliptic plane. The equatorial plane is the flat surface inside the Earth that would join up all the points on the equator. The ecliptic is the apparent trace of the sun through space as the Earth completes its annual orbit of the sun. The offset between the equatorial and ecliptic planes is called the obliquity of the ecliptic. Through time, obliquity changes from 22.1° to 24.5° and back again every 41 ka. Increased obliquity increases the amplitude of the seasonal cycle because both hemispheres experience higher insolation in summer and lower insolation in winter.

The orientation of the Earth relative to the sun defines the seasons, so winter occurs when your hemisphere is pointing away from the sun; and summer when you are pointing towards the sun. For an elliptic orbit, perihelion is the point at which the Earth most closely approaches the sun and aphelion is when the Earth is furthest from the sun. At the moment, the northern hemisphere points away from the sun at perihelion, so winters are relatively warm. At aphelion the Earth is pointing towards the sun, so summers are relatively cool. About 11 ka ago the situation was reversed, summers were at perihelion, so relatively hot, and winters at aphelion, so relatively cold. This phenomenon is called precession, and it occurs on 19 ka and 21 ka cycles.

So, we can take the astronomical cycles and build the gain and phase model (Figure 21). The correlation with the $\delta^{18}O$ record is r = 0.71 indicating that almost three-quarters of the oxygen isotope variation can be attributed to astronomical factors.

CO_2 and temperature from ice

Take another look at the foraminiferal $\delta^{18}O$ record (Figure 17) and notice that the present day represents one of the 100 ka $\delta^{18}O$ minimums (i.e. temperature maximums) and the last 20 ka or so have seen a rapid change from glacial to interglacial conditions. The upshot is that, within a few tens of metres, sea level is as high today as it has ever been within the last 800 ka or so. We are currently close to peak interglacial conditions with expectation that the world should descend back to glacial conditions over the next 10 ka. Similar isotope temperature records have been constructed from ice cores drilled into the polar ice caps (e.g. the Vostok core from Antarctica). Ice records come with additional

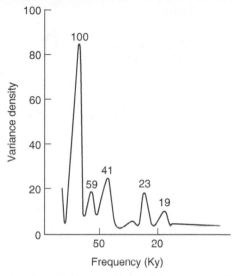

21. **Gain and phase analysis of a consensus Foraminifera $\delta^{18}O$ record showing variance density peaks corresponding to Milankovitch periodicities.**

information because we can reconstruct atmospheric CO_2 composition from the amount of CO_2 dissolved in the ice. These records reveal a very close relationship between temperature and atmospheric CO_2 levels (Figure 22).

Clearly, there are natural fluctuations in atmospheric CO_2 levels that either drive glacial–interglacial cycles or are a response to those cycles. This natural variation can be a source of confusion when it comes to the consideration of anthropogenic emissions of CO_2 to the atmosphere. Let's look at the CO_2 record in a little more detail. Peak interglacials over the past half million years have seen atmospheric CO_2 levels of about 280 parts per million (ppm). Glacials are characterized by atmospheric CO_2 of about 180 ppm. So, a shift of 100 ppm (that's 0.01 per cent) between glacial and interglacial seems to have been a stable climatic oscillation over hundreds of thousands of years.

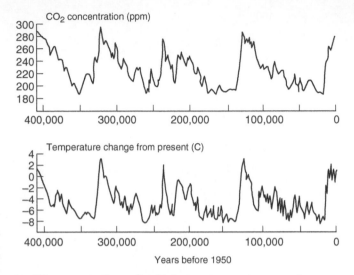

22. CO_2 concentration and stable isotope temperature record from the Vostok ice core showing CO_2 oscillations associated with glacial–interglacial cycles.

We already saw (Chapter 2) the Suess effect in which atmospheric ^{14}C levels were diluted since the industrial revolution by burning oil and coal to introduce 'dead' ^{14}C-free carbon into the atmosphere as CO_2. Now we are seeing a similar effect in atmospheric CO_2. Essentially, carbon that has been stored in the geological record for hundreds of millions of years as coal, oil, and gas has been released into the atmosphere over only a few decades. Moreover, we have done this at a time when CO_2 levels were already rising naturally in the transition from glacial to interglacial. In the last couple of years, atmospheric CO_2 levels have exceeded 400 ppm. So, right at the time when CO_2 levels were hitting the maximum in the 100 ppm oscillation, we have added another 100 ppm to atmospheric CO_2. The figure of 100 ppm doesn't sound much but it is identical to the variation that has accompanied the shift from glacial to interglacial climate extremes. The rate of rise of

CO_2 from glacial to interglacial in the natural cycle looks rapid in the ice core record but it was still achieved over about 20 ka. Mankind has added another 100 ppm slug of CO_2 to the atmosphere over a little more than a century—the size and speed of this CO_2 addition is why climate scientists are so concerned that fossil fuel burning may be making irreversible changes to Earth's climate.

If we look further back into the geological record CO_2 levels much higher than 400 ppm are inferred but it is doubtful that the rate of increase was ever so rapid. It seems that the Earth system has ways of modulating atmospheric CO_2 and temperature over geological time. For example, there is good evidence from sedimentary rocks in the geological record that there has been liquid water at the Earth's surface for more than three billion years, which contrasts greatly with our planetary neighbours, icy Mars and roasting Venus. At issue is probably not whether the Earth can cope with human activity but rather whether people themselves will be the victims of human-induced climate change.

Clumped isotopes

Until now, I have rather glossed over one of the limitations of $\delta^{18}O$ thermometry—the $\delta^{18}O$ technique depends on an isotope exchange reaction that we describe as *heterogeneous*, that is to say that it depends on redistribution of ^{16}O and ^{18}O between two different phases, water and carbonate mineral, i.e.:

$$\tfrac{1}{3}CaC^{16}O + H_2^{18}O = \tfrac{1}{3}CaC^{18}O + H_2^{16}O$$

To constrain this system fully, we need to know the $\delta^{18}O$ of *both* the carbonate mineral *and* the water it grew from. In the geological record we usually have the carbonate but are rarely able to constrain the fluid other than by assumption. The $\delta^{18}O$ thermometer is useful in marine samples because we can assume the $\delta^{18}O$ composition of seawater but it is less definitive in non-marine settings.

A new isotopic thermometer is emerging that is based on the tendency of rare isotopes to clump into molecules at low temperatures and the observation that clumping reduces at higher temperatures. Isotopic clumping was predicted in a series of scientific papers written by Harold Urey in the 1930s exploring the quantum mechanical theory of isotopic effects on the vibrational energies of molecules. Clumping occurs because substitution of a heavy isotope (e.g. ^2D into a hydrogen (H_2) molecule) reduces the molecular vibrational frequency and lowers the so-called 'zero-point energy' (Figure 23). So, D_2 is more stable than HD which is in turn more stable than H_2. Molecules containing more than one rare heavy isotope, so-called 'doubly substituted isotopologues' are even more energetically favoured. While we have known about the thermodynamic prediction of clumping for nearly a century, it is only in the last few years that we have had mass spectrometers capable of resolving the tiny difference in molecular isotopic anatomy that clumping causes.

This clumped isotope phenomenon in carbonates is governed by a *homogeneous* exchange reaction that involves only the carbonate phase, i.e.:

$$Ca^{13}C^{16}O_3 + Ca^{12}C^{18}O^{16}O_2 = Ca^{12}C^{16}O_3 + Ca^{13}C^{18}O^{16}O_2$$

The $Ca^{13}C^{18}O^{16}O_2$ molecule is the clumped species or doubly substituted isotopologue because it incorporates the rare isotopes of both the constituent elements, ^{13}C and ^{18}O. Decreasing temperature encourages clumping, driving the above reaction to the right hand side. Clumped isotope compositions are measured by reacting carbonate minerals with phosphoric acid to liberate CO_2 which is introduced into a mass spectrometer similar to those used to measure $\delta^{18}O$ and $\delta^{13}C$. Such instruments are not able to resolve the various combinations of isotopes that

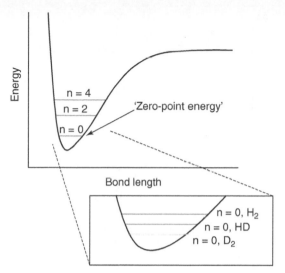

23. Bond energies for different quantum levels (n). Inset shows zero point energies for H_2, HD, and D_2 illustrating that rare isotope substitutions create lower energy (i.e. more stable) molecules. N = 0 is the so-called 'zero-point energy'.

form peaks at the cardinal (i.e. whole number) masses. Instead, we measure all the peaks that occur at mass 47 and compare the sample to a CO_2 gas that has been heated to 1,000°C to remove any clumping. Mass 47 will capture a variety of clumped species ($^{13}C^{18}O^{16}O$, $^{13}C^{17}O^{17}O$, $^{12}C^{17}O^{18}O$). Differences between the sample and the unclumped (or stochastic) composition are expressed as a parameter, Δ_{47}, variously referred to by clumped isotope practitioners as 'big delta 47', 'capital delta 47', or just 'cap 47'.

By growing carbonate minerals in the laboratory under controlled temperature conditions scientists at the California Institute of Technology (CalTech) led by John Eiler first demonstrated the temperature dependence of Δ_{47} and confirmed that many natural

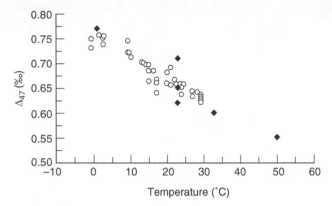

24. **Relationship between mass 47 isotopologues (Δ_{47}) and temperature of carbonate formation. Solid diamonds are experimental calibration data and circles are Foraminifera of known precipitation temperature.**

carbonates plot along the Δ_{47} vs. temperature calibration curve (Figure 24).

One of the more accessible applications of the Δ_{47} thermometer so far has been the work of Rob Eagle and colleagues in Eiler's group. Eagle measured Δ_{47} in the tooth enamel of a range of modern animals. Warm-blooded mammals (a white rhinoceros and an Indian elephant) gave Δ_{47} temperatures within analytical uncertainty of the expected body temperature of 37°C. By contrast, reptiles (Nile crocodile and American alligator) gave Δ_{47} temperatures of about 27°C consistent with their ectothermic (cool-blooded) nature. Two sand tiger sharks, whose body temperature is controlled by their environment, gave Δ_{47} values consistent with the water temperature from which they were caught. Thus, tooth enamel Δ_{47} seems capable of reconstructing animal body temperature.

Eagle and colleagues then measured Δ_{47} in the tooth enamel of five dinosaur fossils. The resultant Δ_{47} temperatures were between

32.4 and 38.2°C with measurement errors of 1–4°C. Thus, these dinosaurs were endothermic (warm-blooded) and endothermism appears to have arisen in the evolutionary progression, from reptiles to birds, in animals whose physical characteristics were still substantially reptilian.

A new sub-discipline of clumped isotope geochemistry, or as John Eiler has called it 'isotomics', is emerging and will have applications far beyond carbonate palaeothermometry. The CalTech group has already demonstrated the temperature dependence of clumping in methane and the future looks very exciting. However, the measurements needed to quantify clumping are challenging. Eiler's laboratory has pioneered the use of high-resolution mass spectrometry using a new instrument known as the 253 Ultra. The ability of a mass spectrometer to resolve molecules with similar mass is known as resolving power. As mentioned in the context of Δ_{47}, 'normal' stable isotope mass spectrometers, with resolving power of around 300, are unable to distinguish molecules with the same cardinal mass. We sacrifice resolving power to allow high-precision isotope ratios. The CalTech prototype 253 Ultra has resolving power in excess of 25,000 but retains isotope ratio measurement capability and the first production model 253 Ultra (now in my lab) has achieved resolving power greater than 40,000. This overcomes the limitations of the conventional mass spectrometers and allows us to recognize and separate molecular peaks down to a fraction of an atomic mass unit. As more laboratories acquire 253 Ultra or equivalent instruments the era of 'isotomics' looks set to revolutionize the application of stable isotopes.

Chapter 8
Scratching the surface with cosmogenic isotopes

Cosmogenic ray interactions

Previously, I described how ^{14}C is produced in the Earth's atmosphere by the interaction between nitrogen and cosmic ray neutrons that releases a free proton turning $^{14}_{7}$N into $^{14}_{6}$C in a process that we call an 'n-p' reaction, which can be characterized, using the shorthand of nuclear reactions, as:

$$^{14}N(n,p)^{14}C.$$

Because the process is driven by cosmic ray bombardment, we call ^{14}C a 'cosmogenic' isotope. The half-life of ^{14}C is about 5,000 years, so we know that all the ^{14}C on Earth is either cosmogenic or has been created by mankind through nuclear reactors and bombs—no 'primordial' ^{14}C remains because any that originally existed has long since decayed.

^{14}C is not the only cosmogenic isotope; ^{16}O in the atmosphere interacts with cosmic radiation to produce the isotope ^{10}Be (beryllium). It will immediately be apparent that this is more complex than the simple n-p reaction because the transmutation of $^{16}_{8}$O into $^{10}_{4}$Be requires the removal of four protons and three neutrons in response to interaction with a cosmic neutron, i.e.:

$$^{16}O(n,4p3n)^{10}Be.$$

The process by which a high energy cosmic ray particle removes several nucleons is called 'spallation'. ^{10}Be production from ^{16}O is not restricted to the atmosphere but also occurs when cosmic rays impact rock surfaces. It is well-known that the amount of exposure to cosmic radiation associated with an intercontinental flight is roughly similar to that of a chest X-ray. As we will see, this is very dependent on where and how high one is flying but the point is that cosmic rays have sufficient energy to pass right through an aircraft fuselage. Similarly, when cosmic rays hit a rock surface they don't bounce off but penetrate the top 2 or 3 metres (m)—the actual 'attenuation' depth will vary for particles of different energy. Most of the Earth's crust is made of silicate minerals based on bonds between oxygen and silicon. So, the same spallation process that produces ^{10}Be in the atmosphere also occurs in rock surfaces. Similarly, spallation of silicon produces another cosmogenic isotope, ^{26}Al by the reaction:

$$^{28}\text{Si}(n,p2n)^{26}\text{Al}.$$

Just like ^{14}C, ^{10}Be and ^{26}Al have half-lives that are short compared to the age of the Solar System (1.36 million years (Ma) and 700 thousand years (ka), respectively) so any primordial ^{10}Be and ^{26}Al is long gone. In Chapter 11, we will see how these extinct isotopes can be used to develop a chronology for the evolution of the Solar System but, for now, let's concentrate on them as cosmogenic isotopes.

Quartz (SiO_2) is a common mineral in the Earth's crust and presents an ideal target for both ^{10}Be and ^{26}Al by having abundant ^{28}Si and ^{16}O and relatively minor concentration of other elements. However, the rate of production of these cosmogenic isotopes is extremely slow—about five atoms per year per gramme of quartz for ^{10}Be and thirty atoms per year per gramme for ^{26}Al. Furthermore, we have already mentioned that ^{10}Be and ^{26}Al are radioactive, so as these isotopes are created, they are also decaying away.

Leaving aside for a moment the question of how one might go about measuring such tiny amounts of these rare isotopes, it is clear that there is a potential chronometer to measure the length of time over which a rock surface has been exposed to cosmic radiation. If we know the flux of cosmic rays impacting a surface, the rate of production of the cosmogenic isotopes with depth below the rock surface, and the rate of radioactive decay, it should be possible to convert the number of cosmogenic atoms into an exposure age.

In practice, the ratios of cosmogenic ^{10}Be to stable ^9Be and cosmogenic ^{26}Al to stable ^{27}Al are of the order of 10^{-13} to 10^{-15} which demand the exquisite sensitivity of accelerator mass spectrometry (see Chapter 6). Typically, we process 30–50 grammes of quartz to make a single ^{10}Be or paired ^{10}Be and ^{26}Al measurement. Samples are carefully purified to remove Al from mineral impurities that aren't quartz and ^{10}Be from atmospheric contamination. Al and Be are then extracted in cleanroom laboratories using ion chromatography (see Chapter 4) and the resulting pure Be and pure Al samples are packed into metal targets for accelerator mass spectrometry.

Cosmogenic isotope production

Knowing the production rate of cosmogenic isotopes is not a trivial issue. As we saw in Chapter 2 for ^{14}C, production rates vary through time—for ^{14}C we call this the de Vries effect. The other cosmogenic isotopes are similarly affected as changes in solar activity impact the Earth's magnetic field which shields us from much of this galactic cosmic radiation. A simple measure of this phenomenon is found by comparing the levels of cosmogenic isotopes in terrestrial rocks and meteorites. Rocks on Earth which are shielded from much of the cosmic radiation have much lower levels of isotopes like ^{10}Be than have meteorites which, before they arrive on Earth, are exposed to the full force of cosmic radiation. Just as we used trees to demonstrate changes in ^{14}C production, polar scientists have used cores drilled through ice sheets in Antarctica and Greenland to compare ^{10}Be at different depths and thereby reconstruct ^{10}Be production through time. The ^{14}C and ^{10}Be records are closely correlated indicating the common response to changes in the cosmic ray flux.

The Earth's magnetic field is thought to be generated by convective movements in the liquid iron sulphide outer core that sits roughly between 2,800 and 5,000 km below the Earth's surface. However, the field is not static. Anyone who has used a magnetic compass in conjunction with a topographic map will be familiar with the idea that magnetic declination varies with time as the difference between the geographic and magnetic poles slowly changes as the magnetic field shifts. Maps produced by the United Kingdom Ordnance Survey usually carry information about the magnetic declination at the time of map publication and a measurement of the rate of change allowing the navigator to adjust compass bearings accordingly.

More dramatic though are magnetic reversals: times when the Earth's field was aligned in the opposite direction so that the magnetic north pole was located close to the topographic south pole. We know this because tiny grains of magnetic iron-rich minerals align themselves with the Earth's magnetic field as they crystallize from lava or are laid down as sediments. On the ocean floor, which is mostly made of basalt lavas, we can detect 'magnetic stripes' of alternating north-aligned and south-aligned magnetic fields which record these geomagnetic reversals as the ocean floor grew through geological time. The exact nature and duration of magnetic reversals are still hotly debated by geophysicists but their existence is not controversial and it is likely that they were times of heightened exposure to cosmic radiation and of increased cosmogenic isotope production.

The Earth's geomagnetic field resembles that of a bar magnet (i.e. a magnet with north and south poles at either end of the magnet). If you take a bar magnet and scatter it with iron filings, the filings concentrate at the poles and arrange themselves along magnetic flux lines that form arcs between the magnetic poles (Figure 25). The particles that make up the cosmic radiation, mostly protons and alpha particles, are similarly deflected by the Earth's magnetic field. The increased concentration of cosmic radiation at the poles is what we see when we observe the northern and southern lights or *aurora borealis* and *aurora australis*, respectively. The upshot is increased cosmogenic isotope production by about four times at the poles compared to the equator. So, to calculate a cosmogenic isotope exposure age, we need to know the latitude at which the sample was exposed.

Another major control of cosmogenic isotope production rate is altitude. At high altitude, there is simply less shielding from cosmic radiation due to the thinner atmosphere. Compared to sea-level, at 3,000 m production rates are about ten times higher and by 5,000 m rates have increased to about thirty times the sea-level rates.

In detail, cosmogenic isotope production rates are governed by a complex array of effects which are extremely difficult to quantify accurately. As a compromise, we tend to present cosmogenic isotope data relative to a production rate model based on one of a small number of attempts that have been made to come up with an accurate model for production rate. As long as scientists using cosmogenic isotopes are clear about which production rate model was used to calculate ages, we can cross-calibrate studies retrospectively. In recent years a broad consensus emerging from the CRONUS-Earth collaboration has led to a user-friendly online calculator for ^{10}Be-^{26}Al exposure ages that offers a standardized age calculation (<http://hess.ess.washington.edu/>).

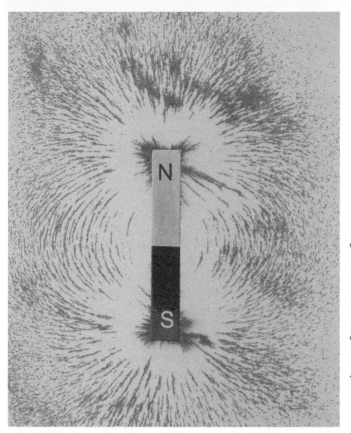

25. Iron filings around a bar magnet to illustrate magnetic flux lines between magnetic poles. The Earth's magnetic field shows a similar distribution of flux.

Exposure dating—age vs. erosion

So, once we have credible cosmogenic isotope production rates, what sorts of scientific problems can we address? Basically, there are two classes of applications, which we can call 'exposure' and 'burial' methodologies. Exposure studies simply measure the accumulation of the cosmogenic nuclide. Such studies are simplest when the cosmogenic nuclide is a stable isotope like ^{3}He and ^{21}Ne. These will just accumulate continuously as the sample is exposed to cosmic radiation. Slightly more complicated are cosmogenic isotopes that are radioactive (e.g. ^{10}Be and ^{26}Al). These isotopes accumulate through exposure but will also be destroyed by radioactive decay. Eventually, the isotopes achieve the condition known as 'secular equilibrium' where production and decay are balanced and no chronological information can be extracted. Secular equilibrium is achieved after three to four half-lives, which make ^{10}Be and ^{26}Al useful up to about three Ma.

Imagine a boulder that has been transported from its place of origin to another place within a glacier—what we call a glacial erratic. While the boulder was deeply covered in ice, it would not have been exposed to cosmic radiation. Its cosmogenic isotopes will only have accumulated since the ice melted. So a cosmogenic isotope exposure age tells us the date at which the glacier retreated, and, by examining multiple erratics from different locations along the course of the glacier, allows us to construct a retreat history for the de-glaciation. Now let's track the same glacier back to its source in the mountains. If we analyse cosmogenic isotopes on the peaks of mountains and we get ages that are much older than we found for the various erratics, then it is fair to conclude that those mountain tops weren't covered by the ice-sheet but stood out as rocky pinnacles above the ice—so called 'nunataks'.

There is a complication though, what if the nunatak itself has eroded since the ice age? The current mountain top would not have been exposed when the peak was a nunatak, so the apparent exposure age would be a complicated mixture of cosmic ray exposure and erosion of the rock surface removing the cosmogenic isotopes. So, all cosmogenic isotope exposure ages can also be interpreted as erosion rates, and it is down to the geomorphologist to provide independent field observations to decide which

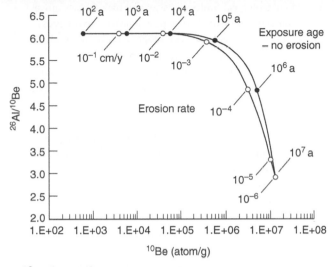

26. ^{26}Al/^{10}Be vs. ^{10}Be 'banana' diagram—upper curve represents exposure age assuming no erosion, lower curve represents erosion rate assuming continuous exposure.

interpretation is preferable. Some additional constraints can be achieved by combining more than one cosmogenic isotope. It is common to combine ^{10}Be and ^{26}Al measurements on the same sample. Continuous exposure and steady-state erosion have subtly different trajectories on plots of ^{26}Al/^{10}Be vs. ^{10}Be—known colloquially as 'banana plots' (Figure 26). Measurement precision is rarely adequate to distinguish the two end-member models but these diagrams have a 'banana' defined by the area between the continuous exposure and steady-state erosion trajectories. Scenarios in which exposure and/or erosion have been more complicated than continuous exposure or steady-state erosion yield ^{10}Be and ^{26}Al that lie outside the 'banana' or 'island of steady-state erosion'.

Cosmogenic isotopes are also being used extensively to recreate the seismic histories of tectonically active areas. Earthquakes occur when geological faults give way and rock masses move. A major earthquake is likely to expose new rock to the Earth's surface. If the field geologist can identify rocks in a fault zone that (s)he is confident were brought to the surface in an earthquake, then a cosmogenic isotope exposure age would date the fault—providing, of course, that subsequent erosion can be ruled out or quantified.

Precarious rocks are rock outcrops that could reasonably be expected to topple if subjected to a significant earthquake. Dating the exposed surface of precarious rocks with cosmogenic isotopes can reveal the amount of time that has elapsed since the last earthquake of a magnitude that would have toppled the rock. Constructing records of seismic history is not merely of academic interest; some of the world's seismically active areas are also highly populated and developed. As we saw from the 2011 Japanese earthquake, tsunami, and Fukushima Daiichi Nuclear Power Plant accident, those areas are very vulnerable to seismic hazard and knowledge of past seismic activity can be usefully incorporated into the risk assessment process when planning major civil engineering projects like nuclear power plants and dams.

Burial dating

Burial methodologies using cosmogenic isotopes work in situations where a rock was previously exposed to cosmic rays but is now located in a situation where it is shielded. One example could be a sedimentary rock deposited in a cave by a river which is gradually cutting down through solid rock to increase the size of the cave. Sedimentary grains which were exposed to cosmic radiation for long periods of time will have developed the secular equilibrium $^{10}Be/^{26}Al$ value. Once the sediment is shielded from cosmic rays, the ^{10}Be and ^{26}Al will continue to decay but no new cosmogenic ^{10}Be and ^{26}Al will be produced. The half-life of ^{10}Be (1.39×10^6 years) is roughly double that of ^{26}Al (7.17×10^5 years). So after one ^{26}Al half-life, half the original ^{26}Al will remain but only about a quarter of the original ^{10}Be will have decayed, and the ratio ^{10}Be/^{26}Al will have doubled compared to the secular equilibrium value. Often, sediments in cave systems will be distributed on a series of terraces abandoned as the river responsible for depositing the sediment has cut down through the rocks. Comparing burial ages for different terraces allows reconstruction of the erosion history of the cave system.

Another burial application of cosmogenic isotopes has allowed insights into the age of sand dunes and the movement of sand grains around the desert. Sand grains that were originally exposed to cosmic radiation become buried and shielded as they are incorporated into dunes. A study of sands from the Namib Desert of Namibia compared abundances of ^{26}Al, ^{10}Be, and non-radioactive ^{21}Ne, and showed that sand dunes have persisted in this desert for over one million years despite major changes in climate. Furthermore, the study showed that it takes at least a million years for sand to cross the 400 km from the southern edge of the desert to its northern margin.

Increasing access to the sensitive accelerator mass spectrometers (^{10}Be, ^{26}Al, and ^{36}Cl) and noble gas instruments (^{3}He and ^{21}Ne) required to measure the rare cosmogenic nuclides is allowing Earth scientists to exploit these isotopes to place quantitative constraints on the exposure and burial histories of geological formations, and thereby assemble more robust reconstructions of such parameters.

Chapter 9
Uranium, thorium, and their daughters

Natural decay series

All isotopes heavier than ^{208}Pb (lead) are radioactive although
bismuth-209, previously thought to be the heaviest stable isotope,
was only discovered to be radioactive in the early years of the
21st century—it has a half-life of over 10^{19} years. However, many
more of these radioactive isotopes have half-lives that are long
compared to the age of our Solar System. These include the
common isotopes of uranium and thorium ^{235}U, ^{238}U, and ^{232}Th,
which have half-lives of 703.8 Ma, 4.468 Ga, and 14.05 Ga,
respectively, where Ma is a million (10^6) years and Ga is a billion
(10^9) years. As we will see shortly, the half-life of ^{238}U is similar to
the age of our Solar System, so about half the ^{238}U that ever existed
has now decayed to other elements. The half-life of ^{235}U is much
shorter and only about 1 per cent of the original ^{235}U remains. By
contrast, Th with its long half-life retains about 85 per cent of the
original (or primordial) ^{232}Th. The principal modes of radioactive
decay are α and β (Chapter 1). $^{238}_{92}$U decays by α-emission to
$^{234}_{90}$Th but ^{234}Th is heavier than ^{208}Pb so will itself be radioactive.
Actually, $^{234}_{90}$Th is a β-emitter, so it decays to $^{234}_{91}$Pa (protactinium)
which itself decays by β to $^{234}_{92}$U which, in turn, decays by α to
$^{230}_{90}$Th. This process continues until a stable lead isotope is
produced—^{238}U finally produces ^{206}Pb, ^{235}U ends up as ^{207}Pb, and

^{232}Th makes ^{208}Pb—and these are known as 'natural decay series' (Figure 27).

Now let's think of a rock made up of crystals that contain U and Pb. The Pb in that rock will be a mixture of the (primordial) Pb incorporated into the rock when it formed and the (radiogenic) Pb that has built up from the decay of U and Th. If we have a series of rocks that all formed at the same time in equilibrium with the same primordial Pb but with different U/Pb ratios, then at any point in the future they will have evolved different, but predictable, ^{206}Pb/^{204}Pb and ^{207}Pb/^{204}Pb. ^{204}Pb is not produced by radioactive decay, so is entirely primordial and it is used as the denominator in these ratios simply because it is much easier to measure isotope ratios accurately than it is to determine individual isotope abundances. On a diagram of ^{207}Pb/^{204}Pb vs. ^{206}Pb/^{204}Pb, samples of the same age but different original U/Pb will fall on a straight line that we call an isochron.

The age of the Earth

This isochron approach was the basis for the first isotopic determination of the age of the Earth. Previously, estimations of the Earth's age had been based on rather less robust reasoning. In the 17th century, biblical scholars like Archbishop James Ussher and John Lightfoot tried to use the Christian Old Testament to work out the date of creation. Ussher worked out that the Earth began at midnight on Sunday, 22 October 4001 BC; Lightfoot obtained a similar result—9 a.m. on 23 September 3928 BC. Today we might describe these observations as remarkably precise but disappointingly inaccurate!

In the 18th and 19th centuries, the age of the Earth debate was led from Scotland, although there were undoubtedly contributors from mainland Europe and my perspective is surely skewed in favour of my adopted home. James Hutton, a leading figure in the

Isotopes

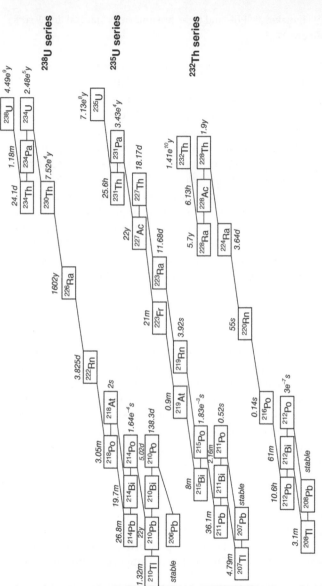

27. **The natural decay series of ^{238}U, ^{235}U, and ^{232}Th.**

Scottish Enlightenment, recognized the gradual pace of geological processes, observing that: 'The result, therefore of our present enquiry is that we find no vestige of a beginning, no prospect of an end.' Across in Glasgow a century later, Lord Kelvin took a different view. He tried to calculate the age of the Earth by modelling the cooling of the planet and came up with an age between twenty Ma and 400 Ma—not especially precise but getting more accurate.

Of course, missing from Kelvin's calculations was the heat generated by radioactive decay—the phenomenon wasn't to be properly discovered for another fifty years or so; and it would be another half century until radioactive isotopes would be used to demonstrate how the very same isotopes have sustained the Earth's internal heat for 4.5 billion years.

In 1955, Claire Patterson working at the California Institute of Technology (CalTech) measured the Pb isotope composition of four different meteorites: an iron meteorite and representatives of the so-called chondrite and basaltic achondrite meteorite classes. These fell on a straight line on a plot of $^{207}Pb/^{204}Pb$ vs. $^{206}Pb/^{204}Pb$, which represents a 4.55 Ga isochron (Figure 28). If these

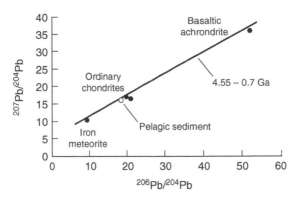

28. **Pb-Pb isochron diagram that first established the age of the Solar System.**

meteorites represent material from the initial Solar System then the suggestion is that the Solar System 'began' 4.55 Ga ago. Patterson also analysed a fine-grained deep sea sediment in the hope that this would be a good representation of the whole Earth. That sediment also plotted on the 4.55 Ga isochron suggesting that the Earth was also formed at the same time as the meteorites and thereby establishing the first really definitive evidence that the Earth is billions of years old. In retrospect, the deep sea sediment result appears to have been something of a fluke. Very few subsequent measurements of similar samples plot on the 4.55 Ga isochron and isotope geochemists now believe that the Pb isotope evolution of the Earth is complicated and that it is unlikely that any single rock exposed at the Earth's surface would be representative of the whole Earth. Notwithstanding, we will see in Chapter 11 that there is compelling evidence that the Earth was already a differentiated planet within a few million years of the initial formation of the Solar System.

U-Pb geochronology

Nowadays, U-Pb dating is well-established as an accurate and precise methodology for geochronology. The method depends on a few minerals, of which zircon is the most common, that crystallize containing high abundances of U but trivial amounts of Pb. Thus, practically all the Pb in a zircon crystal is radiogenic and there is no need to assume an initial Pb isotope composition—actually in practice this is routinely checked by monitoring, and if necessary correcting for, non-radiogenic ^{204}Pb. On a plot of $^{206}Pb/^{238}U$ vs. $^{207}Pb/^{235}U$ (Figure 29) there is a single curve upon which minerals that have retained all the Pb derived from U decay should plot. This 'concordia' represents the curve along which the ^{238}U-^{206}Pb and ^{235}U-^{207}Pb natural series yield the same age.

The U-Pb geochronometer depends on the ability of the host mineral to retain all the intermediate isotopes in the natural decay series from U to stable Pb. This isn't a trivial requirement since

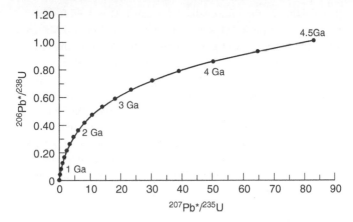

29. U-Pb conchordia plot showing evolution of radiogenic lead through time. The asterisks indicate radiogenic ^{206}Pb and ^{207}Pb implying that any primordial Pb has been removed by correction using non-radiogenic ^{204}Pb.

it requires mineral lattice sites that readily accommodated U to hold onto intermediate elements with different ionic radius and charge state to the original U. Furthermore, the energy released during (particularly α) radioactive decay also causes damage to the mineral lattice that will make these crystals yet more leaky for decay series isotopes. In practice, geochronologists have devised methods of physical and acid chemical abrasion of crystals that are able to remove damaged areas of mineral grains that might yield false geochronological information. The upshot is that the modern U-Pb geochronology laboratory is able to date rocks from the deep-time geological record (i.e. 100 Ma to 100 Ga) with precisions of a few tens of thousands of years.

A good example of U-Pb geochronology in action is dating of the boundary between the Cretaceous and Palaeogene geological periods. Geologists divide the rock record into periods, which are themselves composed of epochs, and which are further divided into ages. The Cretaceous period (abbreviated to 'K' after

'Kreide' the German for chalk and because 'C' is used for the Carboniferous period) lasted from about 145 Ma to 66 Ma ago while the Palaeogene (Pg) spans 66 Ma to 23 Ma, so the K-Pg boundary occurs at 66 Ma—you may also see this boundary referred to as K-T where T is for Tertiary. Geologists used to, and informally often still do, divide the last 66 Ma of geological time into Tertiary (66–2.6 Ma) and Quaternary (<2.6 Ma) sub-eras. Nowadays, it is considered best practice to divide the Tertiary into the Palaeogene (66–23 Ma) and Neogene (23–2.6 Ma) Periods; together the Palaeogene, Neogene, and Quaternary Periods constitute the Cenozoic Era—eras being the highest level of geological division.

The K-Pg boundary, let's call it the KPB, has long fascinated geologists because it is associated with a mass extinction in which 70 per cent of living species, including the last dinosaurs, died out. Over the decades there have been many theories as to why the dinosaurs became extinct and what might have triggered this and other mass extinctions in the geological record. In the early 1980s, physicist Luis Alvarez, his geologist son Walter, and their co-workers published evidence that the KTB was associated with some sort of extraterrestrial impact. Using analytical equipment that was in those days firmly the preserve of physicists and well beyond what was usually available to geology, the Alvarez team discovered anomalously high levels of the element iridium (Ir) at the KTB. Ir is a very rare element in the Earth's crust. Levels at the KTB are still only a few parts per billion (i.e. mg per ton) but even these levels constitute a significant positive anomaly in Ir concentration. Meteorites are greatly enriched in Ir compared to terrestrial rocks, so the team surmised that a massive extraterrestrial impact could have deposited the iridium layer. Moreover, the effect on the Earth's climate could have been sufficient to make the planet uninhabitable for many species.

Subsequently, it emerged that the Ir layer is present around the world in geological sections that contain the KPB. Eventually, a structure was discovered on the Yucatán Peninsula in Mexico that has been interpreted as the impact crater formed by the KTB impact; we now know this as the Chicxulub structure. Proving a direct connection between Chicxulub and the mass extinction will always be problematic because it requires resolution of an instantaneous impact event in geological records which accumulate gradually over millions of years. Notwithstanding, the KTB has brought out the best in geochronology.

A good KTB rock record is preserved in the Hell Creek area of NE Montana. This section is mostly mudstones and sandstones but also includes some low grade coal (lignite) beds and thin horizons of weathered volcanic rocks called bentonites. Critically, the higher of the bentonite layers contains zircons. Fifteen zircons that were chemically abraded to improve the chances of getting a good U-Pb date gave an age of 65.988 +/−0.074 Ma; so the horizon is dated to 66 Ma with a precision of 74,000 years. This bentonite, however, is about 20 metres (m) above the KTB iridium anomaly which must, therefore, be older than 65.988 Ma. A second bentonite is located only a few centimetres above the highest Ir concentration and 5 cm above the last occurrence of Cretaceous fossil pollen. Unfortunately, this bentonite doesn't contain zircons but it does have potassium (K)-rich feldspars called sanidine which can't be dated using U-Pb but are amenable to argon-argon (Ar-Ar) dating.

The Ar-Ar method is based on the decay of ^{40}K to ^{40}Ar; to cut a long story short, it turns out that the best way to quantify K is to irradiate the sample in a nuclear reactor and transmute a proportion of the ^{39}K into ^{39}Ar. By monitoring standards co-irradiated with the samples it is possible to work out the efficiency at which ^{39}Ar is produced within the reactor and a single high-precision $^{39}Ar/^{40}Ar$ measurement gives ^{40}K and ^{40}Ar concentrations and an age whose precision is comparable to the U-Pb method.

KPB deposits around Chicxulub contain tektites, which are gravel-sized fragments of glass produced by melting terrestrial rocks due to the shock of an extra-terrestrial impact. Tektites can be Ar-Ar dated and fourteen Chicxulub tektites from Haiti yielded an average Ar-Ar age of 66.038 +/-0.025 Ma. Sanidines from the bentonite close to the Ir layer gave an average age of 66.011 +/-0.011 Ma. The ages overlap within the highest precision of which geochronology is currently capable and, while +/-20,000 years is still a very long time compared to the instantaneous impact event, it seems that the geological record is trying to tell us that the KPB extinction was triggered by the Chicxulub impact.

However, the detailed story may be a little more complicated. At the time of the Chicxulub impact the Earth was already in trauma. Over in what is now India, which at the time was located in the Indian Ocean roughly between where the Seychelles and Diego Garcia are now, a massive series of volcanic eruptions was underway. The Deccan Traps flood basalt eruptions undoubtedly would have impacted on Earth's climate system. So, our best guess is that towards the end of the Cretaceous, ecosystems were already approaching tipping points due to intense and voluminous Deccan volcanic emissions. Maybe Chicxulub was the straw that broke the camel's back and triggered the final decent into mass extinction?

U-series dating methods

One aspect of the natural decay series that acts in favour of the preservation of accurate age information is the fact that most of the intermediate isotopes are short-lived. For example, in both the U series the radon (Rn) isotopes, which might be expected to diffuse readily out of a mineral, have half-lives of only seconds or days, too short to allow significant losses. Some decay series isotopes though do have significantly long half-lives which offer the potential to be geochronometers in their own right. Of these, ^{234}U ($t_{1/2}$ = 246,000 years), ^{230}Th (75,000 years), ^{231}Pa (33,000 years), ^{226}Ra (1,600 years), and ^{210}Pb (22 years) are the most commonly utilized.

These techniques depend on the tendency of natural decay series to evolve towards a state of 'secular equilibrium' in which the activity of all species in the decay series is equal.

I.e., for species x, y, and z at secular equilibrium:

$$n_x\lambda_x = n_y\lambda_y = n_z\lambda_z = 1,$$

where n represents the number of atoms and λ is the decay constant which is itself just another way of expressing the half-life of a radioactive isotope ($t_{1/2}$) in which $\lambda = \ln(2)/t_{1/2}$ and $\ln(2)$ represents the natural logarithm of 2 (c.0.693) and simply characterizes the exponential nature of nuclear decay. So, at secular equilibrium, isotopes with long half-lives (i.e. small decay constants) will have large numbers of atoms whereas short-lived isotopes (high decay constants) will only constitute a relatively small number of atoms. Since decay constants vary by several orders of magnitude, so will the numbers of atoms of each isotope in the equilibrium decay series. Geochronological applications of natural decay series depend upon some process disrupting the natural decay series to introduce either a deficiency or an excess of an isotope in the series. The decay series will then gradually return to secular equilibrium and the geochronometer relies on measuring the extent to which equilibrium has been approached.

One of the most successful decay series methods has been using ^{234}U-^{230}Th to date marine carbonates like fossil corals. Early attempts to measure ^{234}U/^{238}U in seawater gave the surprising result that seawater is enriched in ^{234}U over ^{238}U by about 15 per cent above the secular equilibrium value. This was unexpected because chemical processes are unlikely to discriminate (fractionate) between two isotopes of the same element with such a small relative mass difference. Moreover, it was already known that the ratio of ^{235}U/^{238}U is constant in natural materials. However, ^{234}U is the great-granddaughter of ^{238}U produced via α-decay to ^{234}Th and two β-decays through ^{234}Pa. The ^{234}U is therefore located in mineral lattice sites that are potentially damaged by the radioactive decay energy. In addition, radioactive decay also encourages oxidation of U^{4+}, which is relatively insoluble in water to the relatively soluble U^{6+}. Finally, the recoil energy associated with emission of the α-particle during the ^{238}U to ^{234}Th decay can be sufficient to 'bounce' the ^{234}Th out of the mineral lattice, so that subsequent ^{234}U forms on the mineral surface. All these factors conspire to make ^{234}U more easily weathered and eroded from rocks than ^{238}U. As rocks weather and erosion transports the weathered material in streams and rivers, and ultimately into the oceans, ^{234}U is preferentially transported relative to ^{238}U. The upshot is the 15 per cent enrichment in $^{234}U/^{238}U$ in seawater.

Minerals that crystallize from seawater, for example the carbonate skeletons of corals, inherit this 15 per cent ^{234}U excess, which means they form out of secular equilibrium. Typically, such minerals incorporate negligible ^{230}Th because Th is very insoluble in seawater. So, any ^{230}Th in the mineral represents in-growth from the decay of the excess ^{234}U and the evolution towards secular equilibrium between ^{234}U and ^{230}Th can be used as a chronometer. Combining U-Th dating with the fact that many corals incorporate annual growth bands (similar to tree rings) and that parameters such as $\delta^{18}O$ and ratios of trace elements are sensitive proxies for palaeoclimate has allowed scientists to reconstruct climate records with time resolution that is able to recognize short-term variability in climate over tens of thousands of years. In one such study, our group was able to use corals in Papua New Guinea to sample time segments over the past 120,000 years with a resolution of about three months per sample. Statistical analysis of the data recognized a variability on a timescale of three to seven years similar to the El Niño Southern Oscillation that affects SouthEast Asia and South America. So, rather than being a result of man-made climate change, the El Niño effect appears to be a natural phenomenon. However, our study also noticed that the El Niños of the late 20th century were the strongest of any in the last 120,000 years. Thus, while El Niño may be a natural phenomenon, it is also possible that the effect of El Niño has been amplified by human activities.

Nuclear fission

In Chapter 1, we introduced the concept of spontaneous nuclear fission. In the heaviest elements, in addition to α-decay, occasionally nuclides split into two atoms that sum to slightly less than the atomic mass of the original fissile isotope. The fission products of ^{235}U vary from fission to fission but tend to fall into two groups in the mass range 90 to 100 and 130 to 150, respectively. For instance, well-known ^{235}U fission products, because they constitute significant radiological hazard in the aftermath of a nuclear explosion, are ^{90}Sr and ^{137}Cs. But $90 + 137 = 227$ not 235 which indicates that a few neutrons are not incorporated into the two fission products. Those neutrons can themselves go on to induce fission in other fissile nuclides. If the yield of free neutrons from fission of a particular isotope is statistically greater than one, every fission results in more than one further fission and a nuclear chain reaction is set in motion. Each fission releases energy and the upshot of a chain reaction is the generation of colossal amounts of the energy that used to hold the fissile isotope together. In a nuclear fission bomb that energy is all released in an instant whereas in a nuclear reactor the chain reaction is controlled using materials (e.g. boron and cadmium) that are able to absorb neutrons without undergoing fission themselves. These 'control rods' can be moved in and out of the nuclear fuel to control the rate of the nuclear action, allowing generation of energy without a run away chain reaction.

Establishing a nuclear chain reaction requires a critical mass of the fissile isotope. The critical mass of ^{235}U is 52 kg, which would be a sphere of 17 cm diameter. Natural uranium will not sustain a chain reaction because the present day proportion of ^{235}U (0.72 per cent) is too low—it is ^{235}U rather than the more abundant

^{238}U that is the main source of fissions. To produce a nuclear bomb requires the proportion of ^{235}U to be artificially increased. Nuclear weapons rely on enrichment rates of more than 90 per cent ^{235}U; lower enrichments could support a weapon but would involve such large amounts of uranium that deployment of the weapon would be entirely impractical. Nonetheless, any ^{235}U enrichment above 20 per cent is technically designated as 'weapons grade'. Nuclear reactors use much lower levels of ^{235}U enrichment of only a few per cent and, like natural uranium, will not sustain a nuclear explosion. The explosions seen during the Fukushima Daiichi accident were not due to the uranium oxide nuclear fuel exploding but rather by reaction between the zirconium alloy (zircaloy) used in fuel rod construction and the water used in this type of boiling water reactor. Under normal operating conditions, zircaloy is inert, but when the Fukushima cooling systems failed and the reactor temperatures ran out of control, zircaloy reacted with water to release highly reactive hydrogen gas and it was ignition of the hydrogen that caused the Fukushima explosions.

A natural fission reactor

As mentioned earlier in the chapter the ratio of ^{235}U/^{238}U is constant in all uranium with ^{235}U always constituting 0.72 per cent of total U—well, not quite. In 1972 workers in a French nuclear reprocessing plant measured uranium ore from a deposit at Oklo in Gabon (a former French colony in West Africa) that appeared to be deficient in ^{235}U. After further investigation it was discovered that parts of the uranium deposit were missing significant amounts of ^{235}U—enough for several critical masses. Clearly, something strange had happened at Oklo, and it had happened several times because anomalous levels of ^{235}U were detected at sixteen separate areas in the Oklo and adjacent Okelobondo uranium mines.

The key to what had happened was found in the isotopic composition of other elements—the fission products. Elements like neodymium (Nd) and ruthenium (Ru) at Oklo have isotope compositions that are different from those normally observed for materials on Earth. Moreover, when the discrepancies were compared they matched exactly with the expected yield of those isotopes from the fission of ^{235}U. Apparently, the uranium deposits of Gabon behaved as natural nuclear reactors. Previously, we said that natural uranium had insufficient ^{235}U to sustain a chain reaction. However, the Gabon deposits were formed about 1.7 Ga ago. 1.7 Ga is almost 2.5 half-lives for ^{235}U, so 1.7 Ga ago the proportion of ^{235}U would have been about 3 per cent—not dissimilar to the enriched uranium we now use in nuclear reactors.

Still, simply concentrating enough ^{235}U in one place does not guarantee a natural reactor. ^{235}U fission occurs when ^{235}U interacts with a slow-moving neutron, but the neutrons released by fission are fast-moving. The chain reaction will only work if the fission-generated neutrons are slowed down by a 'moderator'. Power reactors typically use graphite or water as the moderator. Similarly, the Oklo natural reactors would have needed to be moderated—how could this have been achieved naturally? The answer seems to be that the ore deposits are hosted in sandstone rocks which are porous to water. Rain water at the surface soaks into the sandstone to form a saturated layer of rock that geologists call an aquifer. It is into such aquifers that people in arid regions drill bore-holes to access groundwater for drinking and irrigation. Similarly, it is these porous rock types that often store oil and gas deposits.

At Oklo, it seems that groundwater was able to penetrate into the uranium deposits where it offered a perfect moderator to allow the operation of the natural nuclear fission reactor. Obviously, much of the energy of fission is released as heat—indeed the boiling water reactor types, like those at Fukushima Daiichi, use this heat to produce steam which drives the turbines that generate electricity. Wouldn't a natural reactor moderated by groundwater simply boil off the water and destroy its own moderator? Well actually, yes. The different isotopes of the noble gas xenon (Xe) would be produced over very different lengths of time in a nuclear reactor—this is well known because Xe isotope fission products significantly affect nuclear reactor operation and it was a failure to properly understand the moderating effect of ^{135}Xe production that was a key contributor to the Chernobyl accident. At Oklo, scientists mathematically modelled the distribution of Xe isotopes and worked out that the reactors cycled. As water entered the uranium deposit, the reactors operated for about thirty minutes until temperatures increased enough to boil off the water. The reactor would then shut down for about 2½ hours until enough groundwater accumulated to restart the fission chain. Amazingly, this eight-cycles-a-day routine seems to have lasted for hundreds of thousands of years.

The geochemists who studied Xe at Oklo used a laser beam to vaporize the minerals that contained Xe and then transported the Xe into a mass spectrometer. In preparation for these experiments they needed to establish where the Xe atoms were located in the rocks. Contrary to expectations, Xe wasn't associated with the U-rich minerals in which it would have been produced. Rather, the Xe was concentrated in aluminium phosphate minerals. It seems that the uranium-rich minerals were not able to retain the Xe fission product as temperatures in the reactor rose but, at the same time, the Xe was trapped in the aluminium phosphates as they grew in response to the increase in temperature. While the individual uranium minerals didn't hold on to their fission products, the overall deposit remained self-sealing for nearly two billion years. Clearly, there are important lessons here for the disposal of spent nuclear fuel.

The Oklo deposits record a quite remarkable coming together of circumstances which conspired to create a natural nuclear reactor which appears to have operated in a stable manner for a considerable length of time. But Oklo might be telling us something more fundamental about the evolution of our planet. Many uranium mineral deposits are formed because reduced U^{4+} is much less water-soluble than oxidized U^{6+}. Often, uranium concentrates when oxidizing fluids encounter reducing conditions and U^{6+} in solution is reduced to insoluble U^{4+}, which precipitates and builds up. So, significant concentrations of uranium only occur because the Earth's atmosphere is oxidizing. But, it wasn't always so—prior to about 2.5 Ga ago, geochemists believe that the atmosphere was far more reducing. Remember that ^{235}U would have been a much higher proportion of total uranium prior to 2.5 Ga. So, it is probably a good thing that significant concentrations of uranium were not geologically feasible when fissile ^{235}U was so abundant. Or, is it that the Earth only survived its initial concentration of explosive ^{235}U because the atmosphere became oxidized only after ^{235}U levels had decayed to only a few per cent of total uranium? One can only speculate on the fate of a planet whose ^{235}U and atmospheric oxygen were not so finely balanced.

Chapter 10
Probing the Earth with isotopes

Earth's internal structure

Over four billion years of active geology has made Earth a very differentiated planet. The outer few tens of kilometres (km) is known as the crust, and can be divided into continental and oceanic crust. The continental crust includes the exposed landmasses and the adjacent continental shelves, which represent parts of the continents that are currently flooded by the seas to a depth of about 150 metres (m). The continental crust is enriched in elements such as silicon and aluminium which make it low in density compared to the whole Earth. The upshot is that once continental crust is formed it becomes an almost permanent structure as the outer layer of the Earth. Typically, continental crust is about 40 km thick although there are areas of both thicker and thinner continent.

Oceanic crust is predominately composed of basalt rocks, which are relatively rich in iron and magnesium. These basalts form at mid-ocean ridges, which are chains of volcanoes mostly submerged beneath the Earth's oceans—Iceland is an exception in being a part of the Mid-Atlantic Ridge that rises above sea-level. As the underlying mantle wells up to fill the area vacated as the continents are pulled apart by tectonic forces, decompression of the solid mantle rocks causes them to melt much in the way that popping a champagne cork allows dissolved carbon dioxide gas to separate from the liquid wine. Basalt is dense compared to continental rocks and eventually returns to the mantle in a process that we call subduction. The 'ring of fire' volcanoes around the margin of the Pacific Ocean are a manifestation of subduction in which the oldest parts of the Pacific Ocean crust are being returned to the mantle below. The oldest parts of the Pacific Ocean crust are about 150 million years (Ma) old, with anything older having already disappeared into the mantle via subduction zones. The Atlantic Ocean doesn't have a ring of fire because it is a relatively young ocean which started to form about 60 Ma ago, and its oldest rocks are not yet ready to form subduction zones. Thus, while continental crust persists for billions of years, oceanic crust is a relatively transient (in terms of geological time) phenomenon at the Earth's surface.

Below the crust, geologists divide the sub-surface into the mantle and the core. From the base of the crust to about 2,800 km depth, the Earth is rocky and composed of minerals like olivine and pyroxene that are rich in magnesium, iron, and calcium. At higher and higher pressures these minerals transform into ever denser mineral structures that can be mapped by the speed at which seismic waves generated by earthquakes travel through the Earth. A particularly significant seismic change occurs around about 650 km depth, which we identify as the transition between the upper third of the mantle and the remaining lower mantle. Geophysicists have designated the base of the mantle as the D″ (D double prime) layer which itself demonstrates significant topography that can be mapped with seismic waves. Below D″ the Earth is a metallic alloy of iron and nickel mixed with a small proportion of lighter elements, possibly sulfur or oxygen. From about 2,800 km to about 5,100 km depth the outer core is liquid. We know this because there are seismic waves, so-called S-waves, that can only pass through solids, and the failure of S-waves to be detected in the shadow of the outer core indicates that it is liquid. The remaining 1,200 km or so to the centre of the Earth is also Fe-Ni alloy but is solid metal.

Meteorite models

So, the Earth has separated over geological time into a series of 'reservoirs' and there are no terrestrial rocks that record the earliest history of the planet. However, just as we did for the age of the Earth, we can turn to meteorites to fill in the gaps. In Chapter 1, the rare earth or lanthanide elements were introduced whose chemical behaviour is so similar that they are given a separate box in the periodic table of the elements that we can think of as being squeezed into the main table (Figure 2) between lanthanum (La) and hafnium (Hf). We also noted that $^{147}_{62}Sm$ decays to $^{143}_{60}Nd$ with a half-life of about a hundred million years, so small variations in $^{143}Nd/^{144}Nd$ reflect differences in Sm/Nd integrated over the age of the Solar System—which we assume to have begun with homogenous Sm/Nd.

The group of meteorites we call 'chondrites' display a very narrow range of Sm/Nd, so much so that we need to separate chondrites into their constituent minerals in order to produce isochrons but when this is done we end up with a tightly constrained average Sm/Nd and initial $^{143}Nd/^{144}Nd$. It seems reasonable to assume that the initial Earth had a similar composition. We can take this initial composition and extrapolate through time to form a reference that we call the Chondritic Uniform Reservoir (CHUR), which denotes the Nd isotopic composition of the bulk Earth through geological time (Figure 30). It is unlikely that any Earth reservoir preserves the CHUR composition but variations from CHUR can be very revealing. However, such variations will be small and our ability to exploit the Sm-Nd system relies on our ability to resolve variations in $^{143}Nd/^{144}Nd$ down to the fifth decimal place.

The power of Sm-Nd rests in the predictable behaviour of these two highly similar elements. Few geological processes differentiate between Sm and Nd but an important exception is melting of the

Earth's mantle. Mantle rocks typically contain minerals such as olivine, pyroxene, spinel, and garnet. Unlike say ice, which melts to form water, mixtures of minerals do not melt in the proportions in which they occur in the rock. Rather, they undergo partial melting in which some minerals (e.g. clinopyroxene) melt preferentially leaving a solid residue enriched in refractory minerals (e.g. olivine). We know this from experimentally melting mantle-like rocks in the laboratory, but also because the basalts produced by melting of the mantle are closer in composition to Ca-rich (clino-) pyroxene than to the olivine-rich rocks that dominate the solid pieces (or xenoliths) of mantle that are sometimes transferred to the surface by certain types of volcanic eruptions.

During partial melting, Sm and Nd both show a preference for the liquid melt compared to the solid residue but Nd shows a slightly higher affinity for the melt phase than does Sm. Thus, partial melts of the mantle have lower Sm/Nd than the original mantle whereas the residual mantle left after partial melting has higher Sm/Nd. Low Sm/Nd partial melts will evolve through time to have lower $^{143}Nd/^{144}Nd$ than CHUR and the partially melted residuum will evolve higher $^{143}Nd/^{144}Nd$ than CHUR (Figure 30).

^{87}Rb (rubidium) decays to ^{87}Sr (strontium) with a half-life of about 48.8 Ga—so, less than 10 per cent of primordial ^{87}Rb has decayed in the life-time of the Earth, but this is enough to induce variations in $^{87}Sr/^{86}Sr$ in the Earth's crust and mantle. During partial melting of the mantle, Rb has a higher affinity for the melt phase than Sr, so melts have high Rb/Sr and evolve high $^{87}Sr/^{86}Sr$, and residues have low Rb/Sr and evolve low $^{87}Sr/^{86}Sr$. This means that partial melts should combine low $^{143}Nd/^{144}Nd$ and high $^{87}Sr/^{86}Sr$ and residues from partial melting should combine high $^{143}Nd/^{144}Nd$ and low $^{87}Sr/^{86}Sr$. This being the case, isotope geochemists have found it useful to consider Nd and Sr isotope data together and commonly plot $^{143}Nd/^{144}Nd$ against $^{87}Sr/^{86}Sr$ (Figure 31) for a variety of rocks taking the origin as CHUR for $^{143}Nd/^{144}Nd$ (0.512638) and our best estimate of bulk Earth for

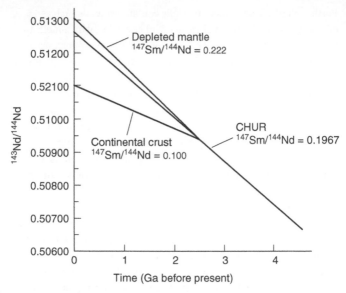

30. **The CHUR model through geological time and projected trajectories for continental crust and depleted mantle formed at 2.5 Ga indicating that continental crust will evolve $^{143}Nd/^{144}Nd$ lower than CHUR and depleted mantle will have higher $^{143}Nd/^{144}Nd$ than CHUR.**

$^{87}Sr/^{86}Sr$ (0.7047). Partial melts (low $^{143}Nd/^{144}Nd$ and high $^{87}Sr/^{86}Sr$) will evolve to plot in the bottom right (or enriched) quadrant whereas residues from melting (high $^{143}Nd/^{144}Nd$ and low $^{87}Sr/^{86}Sr$) should plot in the upper left (or depleted) quadrant.

Continental crust and depleted mantle

By and large our predictions are upheld by data. Continental crustal rocks occupy a wide range of Sr and Nd isotopic values but mostly occupy the enriched quadrant reflecting their ultimate origin as partial melts of the mantle (Figure 31). We expect a big variation because the continents contain materials that ultimately separated

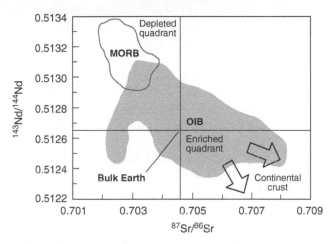

31. $^{143}Nd/^{144}Nd$ vs. $^{87}Sr/^{86}Sr$ for mid-ocean ridge basalts (MORB) and ocean island basalts (OIB).

from the mantle any time since the Earth formed—the oldest crustal rocks exposed at the Earth's surface are about 4 Ga old and will include 4 Ga of rapid ^{87}Sr and sluggish ^{143}Nd accumulation (compared to CHUR), whereas the youngest continental crust is still forming in places like the Andes and has been extracted recently from the mantle and will have mantle-derived low $^{87}Sr/^{86}Sr$ and high $^{143}Nd/^{144}Nd$.

Basalts erupted at mid-ocean ridges occupy a relatively small portion of Figure 31 and are exclusively located in the depleted quadrant. In such young (compared to the half-lives of ^{147}Sm and ^{87}Rb) rocks the high $^{143}Nd/^{144}Nd$, low $^{87}Sr/^{86}Sr$ signature must be inherited from their mantle source. The obvious conclusion is that MORB derive from a mantle source that is complementary to the continental crust. Essentially, we believe that the continents ultimately derive from partial melting of a CHUR-like mantle. However, the amount of mantle melt required to produce the continents is quite small, and this complementary depleted mantle remains capable of yielding basaltic melts when it is decompressed at mid-ocean ridges.

Now, if we estimate the composition and volume of the continental crust and we infer the composition of the complementary depleted mantle from the geochemistry of MORB, then it ought to be possible to work out the amount of the Earth's mantle that has to have had magma extracted to produce the continental crust (as mentioned earlier, the oceanic crust is a transient feature on the billion-year timescale that the continents were constructed). In detail, there are lots of complications that affect what started out as a fairly simple calculation and it is far from clear that we yet have a definitive result. However, it is intriguing that a number of approaches to this problem concluded that only ½ to ⅓ of the whole mantle is required to be depleted to generate the present volume of continental crust. That in turn suggests that the upper mantle (above the 650 km seismic transition zone) might be depleted while the lower mantle (below 650 km) could be relatively pristine. However, I would caution that this is a controversial assertion and there are ways of framing the same problem that come to very different conclusions—such is the fun of doing isotope geochemistry in the almost inaccessible deep Earth.

Ocean Island Basalts

The MORB erupted along the network of mid-ocean ridges are by far the most abundant of oceanic basalts but there are many occurrences of basaltic volcanoes far away from the mid-ocean ridges and the plate-edge subduction zones. These intraplate volcanoes tend to form volcanic islands and we group them as Ocean Island Basalts (OIB). In some cases, OIB volcanoes appear to occur above areas of long-lived anomalously hot mantle so that chains of volcanoes are generated as the ocean plate drifts over a relatively static hot spot in the underlying mantle. Perhaps the best example of a hot spot trace is the Hawaii–Emperor

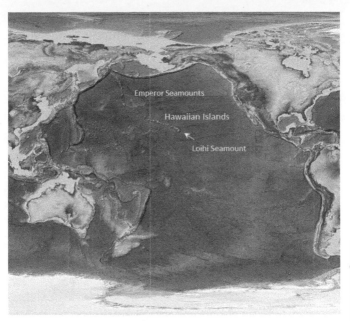

32. A bathymetric map of the western Pacific Ocean showing the Hawaiian island chain and its northward extension into the Emperor seamount chain. Together, these volcanic islands and submarine seamounts are believed to reflect the movement of the Pacific plate over a relatively static deep-seated hot spot or plume in the Earth's mantle.

seamount chain (Figure 32). Geochemists have proposed that a deep-seated thermal anomaly or 'mantle plume' is located beneath Hawaii, specifically this is currently under the Loihi seamount to the southeast of the 'big island' of Hawaii. To the north of Hawaii are the other Hawaiian islands which are thought to have been active volcanoes when they were immediately above the mantle plume. Beyond the northernmost Hawaiian island (Kauai) are the atolls of the Northwestern Hawaiian islands and the Emperor seamounts and guyots, which are the submerged ancestors of Hawaiian mantle plume volcanic activity.

Occasionally, a mantle plume rises beneath a mid-ocean ridge and many geologists believe this to be the reason why the Mid-Atlantic Ridge is exposed above sea level to form the main island of Iceland. In such a situation the decompression melting of the mid-ocean ridge combines with the deep-seated mantle plume thermal anomaly and the resultant enhancement of volcanic activity generates a greater thickness of oceanic crust—the crust beneath Iceland is about 25 km thick compared to typical ocean crust of about 8 km thickness.

Elsewhere, it is less clear that OIB islands can be related to either mid-ocean spreading or mantle plume activity, but nevertheless there are volcanic islands that reveal the isotopic composition of the underlying mantle. The isotopic composition of OIB for $^{87}Sr/^{86}Sr$ and $^{143}Nd/^{144}Nd$ show a much wider range than for MORB. In general, OIB have higher $^{87}Sr/^{86}Sr$ and lower $^{143}Nd/^{144}Nd$ than MORB and some OIB even have higher $^{87}Sr/^{86}Sr$ and lower $^{143}Nd/^{144}Nd$ than CHUR (Figure 31). Clearly, the Earth's mantle is more complicated than simply a differentiation into enriched continental crust and depleted residual mantle. Undoubtedly, the mantle is compositionally heterogeneous; there are many theories as to how heterogeneity has been produced but I think it is fair to say that definitive tests between hypotheses have yet to emerge from geochemistry. What is clear though is that plate tectonics has operated on Earth for up to 4.5 billion years and that has to have involved recycling of crustal materials back into the mantle through subduction zones—some of those materials appear to have preserved high Rb/Sr and low Sm/Nd that have evolved high $^{87}Sr/^{86}Sr$ and low $^{143}Nd/^{144}Nd$ and have remained unmixed from the 'ambient' mantle at a scale length that allows them to be sampled by OIB magmatism. Thirty years ago geologists fiercely debated whether the mantle was homogeneous or heterogeneous; mantle isotope geochemistry hasn't yet elucidated all the details but it has put to rest the initial conundrum; Earth's mantle is compositionally heterogeneous.

Continental basalts

Basaltic lavas erupted through continental crust show more
variation in isotopic composition than the MORB and OIB of the
ocean basins (Figure 33). There is no reason to suggest that the
convecting mantle beneath the continents and oceans is any
different. Over millions of years the same mantle will have found
itself sub-oceanic and sub-continental at various times as the mantle
convected and the continents drifted. So, the wider range of
$^{87}Sr/^{86}Sr$ and $^{143}Nd/^{144}Nd$ in continental basalts (compare
Figures 31 and 33, noting different scales) must be inherited from
the continental plate. As we have seen, the continents are old and
have high Rb/Sr and low Sm/Nd, so have evolved high $^{87}Sr/^{86}Sr$ and
low $^{143}Nd/^{144}Nd$. Continental rocks also tend to have lower melting
temperatures than mantle-derived basalts. For example, basalt lavas

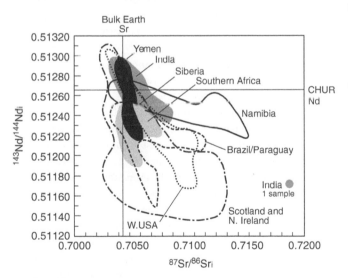

**33. Sr and Nd isotope composition of selected continental basaltic
rocks adjusted to the 'initial' (i) values they would have had when
erupted, i.e. corrected for build-up of radiogenic ^{143}Nd and ^{87}Sr since
the rocks were erupted as volcanic lavas.**

are erupted at about 1,200°C whereas a crustal rock of granitic composition in the presence of water will start to melt at below 700°C. So, it is understandable that basaltic rocks passing through the continental crust will partially melt or assimilate the surrounding crustal rocks thereby taking on 'crustal' isotopic characteristics like high $^{87}Sr/^{86}Sr$ and low $^{143}Nd/^{144}Nd$. Many studies have shown that 'crustal contamination' is a common phenomenon that is generally revealed as increasing $^{87}Sr/^{86}Sr$ and decreasing $^{143}Nd/^{144}Nd$ accompanied by progressively contaminated magmas taking on increasingly crustal characteristics such as increased SiO_2 and Al_2O_3 concentration and decreased MgO content.

However, some continental magmas have high $^{87}Sr/^{86}Sr$ and low $^{143}Nd/^{144}Nd$ but show very little evidence that they have been contaminated with continental crust. For example, some high $^{87}Sr/^{86}Sr$ and low $^{143}Nd/^{144}Nd$ basalts have high concentrations of elements like titanium (Ti), potassium (K), and Sr itself that are too high to be derived from contamination with continental crust.

So, the idea has emerged that there is another source of high $^{87}Sr/^{86}Sr$ and low $^{143}Nd/^{144}Nd$ material whose overall chemical characteristics are mantle-like but which are intimately associated with the continental plates and whose isotopic characteristics are more like continental crust than depleted mantle. We now think that the bottom of the continental crust is not the base of the continental plate. Instead, the continental tectonic plates are much thicker; at least 100 km, maybe even 500 km thick with a mantle keel which we call 'lithospheric mantle'. Lithospheric mantle moves with the continental crust in plate tectonics and so, like the crust, can be isolated from the convecting mantle for billions of years. Furthermore, the mantle xenolith samples transported to the Earth's surface by volcanic activity seem to suggest that lithospheric mantle records processes and episodes that enhance Rb/Sr and reduce Sm/Nd leading to lithospheric mantle having higher $^{87}Sr/^{86}Sr$ and lower $^{143}Nd/^{144}Nd$ than typical mantle rocks and much closer to continental crust.

Mantle geochemists have been arguing about the relative roles of crustal contamination and enriched mantle sources for at least three decades. Different continental volcanic provinces yield evidence for both. So, while there are still many details to be resolved and we have a multitude of questions that have not yet been answered, we cannot underestimate the extent to which isotope data have allowed us to probe the deep Earth. Using isotopes, we have come to understand how our planet has evolved through geological time and we have a refined knowledge of the structure and behaviour of the Earth many tens and hundreds of kilometres beneath our feet.

Chapter 11
Cosmic stopped clocks

To determine the duration of any event—for example, the time it took you to read the last chapter—requires two measurements of time: at the beginning and the end of the event. We are used to committing the first time measurement to memory but we weren't around to do that back in deep geological time. So instead, we have to rely on clocks that stop when the event we are interested in began and combine them with clocks that continued to tick throughout the event. The Canyon Diablo meteorites are fragments of a 300,000-ton nickel-iron meteorite that was responsible for the famous Meteor or Barringer Crater near Flagstaff in Arizona. A sub-sample of Canyon Diablo was one of the meteorites analysed by Patterson to determine the age of the Solar System (Chapter 9). One of the minerals in Canyon Diablo is troilite, which is an iron sulfide chemically similar to pyrite or 'fool's gold'. When troilite crystallizes it incorporates lead but not uranium. So, the lead isotope composition of Canyon Diablo troilite is exactly the same as it was when the meteorite formed. The $^{206}Pb/^{204}Pb$ of Canyon Diablo troilite is about 9 whereas a typical crustal rock has $^{206}Pb/^{204}Pb$ around 18, and the additional ^{206}Pb has been produced by the decay of ^{238}U. So, Canyon Diablo troilite is a stopped clock which records the Pb isotope composition at the start of the Solar System.

Pb ore deposits

The main lead ore mineral is galena which is lead sulfide. Similarly to troilite, galena forms with abundant lead but negligible uranium. So, large lead ore deposits that have concentrated lead from substantial volumes of the Earth's crust provide a series of stopped clocks that document how the Pb isotope composition of the crust has evolved through time (Figure 34).

The most surprising aspect of these combined stopped clocks is that a single U/Pb (μ) ratio for the Earth cannot explain the evolution from an initial Earth with Canyon Diablo Pb to present-day Pb isotope compositions. The simplest possible model involves evolution of Pb in a relatively low U/Pb environment from 4.56 to 3.7 billion years (Ga) followed by evolution in a relatively high U/Pb environment from 3.7 Ga to the present day. Of course, this is just a model and reality likely involves a more complex evolution that is approximated by this two-stage U/Pb model.

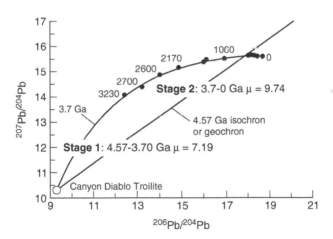

34. Pb isotope compositions of lead ore deposits of different geological age, μ is geochemical shorthand for $^{238}U/^{204}Pb$.

Solar System rubidium-strontium

The next best thing to stopped clocks are clocks that tick very slowly and which started at the same time. ^{87}Rb decays to ^{87}Sr with a half-life of about 48.8 Ga. Over millions of years, materials of the same age but with different $^{87}Rb/^{86}Sr$ (^{86}Sr is simply chosen as one of the stable isotopes of Sr) will evolve ^{87}Sr at a slightly different rate. Samples that start out with the same $^{87}Sr/^{86}Sr$ but different $^{87}Rb/^{86}Sr$ will evolve ^{87}Sr at rates that reflect the initial $^{87}Rb/^{86}Sr$. So, on a plot of $^{87}Sr/^{86}Sr$ against $^{87}Rb/^{86}Sr$, samples that start out on a horizontal line, because they originated with identical $^{87}Sr/^{86}Sr$, gradually evolve onto progressively steeper lines that we call 'isochrons' as higher ^{87}Rb generates more ^{87}Sr.

Basaltic achondrite or eucrite meteorites are a class of meteorite that are similar to the rock type we call basalt on Earth. The eucrites have relatively low, but different, $^{87}Rb/^{86}Sr$ ratios, so they are slowly ticking clocks which tick at slightly different speeds. Thus, while today they have $^{87}Sr/^{86}Sr$ slightly higher than when they originally formed, the difference is quite small, so the uncertainty in calculating their 'initial' $^{87}Sr/^{86}Sr$ is also small.

Eleven basaltic achondrites (Figure 35) were analysed by G.J. Wasserburg and his colleagues from their 'Lunatic Asylum' laboratory (named for their shared obsession with the Moon in the days of the Apollo landings) at the California Institute of Technology. These rocks defined a 4.47 +/–0.24 Ga isochron, within error of Patterson's Pb-Pb 'age of the Earth'. The isochron allows extrapolation to $^{87}Rb/^{86}Sr = 0$—in other words, what a stopped clock would have read—$^{87}Sr/^{86}Sr = 0.69899$. Wasserburg and colleagues called this value BABI for Basaltic Achondrite Best Initial ratio. The observant reader will have noticed that elsewhere in this book $^{87}Sr/^{86}Sr$ values are always greater than 0.7 because radiogenic growth only allows $^{87}Sr/^{86}Sr$ to evolve higher values and we consider BABI to be the initial $^{87}Sr/^{86}Sr$ of the Solar System.

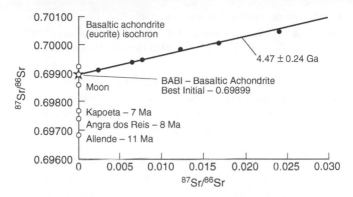

35. Rb-Sr isochron plot for basaltic achondrites (solid circles), Moon rocks and meteorites Kapoeta, Angra dos Reis, and Allende (open circles) have lower initial $^{87}Sr/^{86}Sr$ than the Basaltic Achondrite Best Initial (BABI) $^{87}Sr/^{86}Sr$ that can be converted into ages (Ma) before BABI assuming the rocks evolved in an environment with solar $^{87}Rb/^{86}Sr$.

Rocks from the moon gave initial $^{87}Sr/^{86}Sr$ close to BABI but some meteorites have significantly lower $^{87}Sr/^{86}Sr$. If we assume that these rocks evolved in an environment with the same $^{87}Rb/^{86}Sr$ as the sun, then we can convert this ^{87}Sr deficiency (relative to BABI) into a time before BABI. So-called 'calcium-aluminium inclusions' (CAIs) are the highest temperature minerals in some meteorites that are thought to be the earliest materials to have condensed from the solar nebula. CAIs from the Allende meteorite have $^{87}Sr/^{86}Sr$ about 0.0003 lower than BABI which translates into about 11 million years (Ma). So, within 11 Ma of the formation of the Solar System, the earliest planetary materials were beginning to crystallize.

Extinct isotopes

In Chapter 8, we saw how ^{26}Al is produced by cosmic ray bombardment and its decay can be used to create surface exposure ages. ^{26}Al would have existed in the initial Solar System but only for the first few million years because ^{26}Al has a half-life of about

0.7 Ma. But radioactive ^{26}Al decays to stable ^{26}Mg—could it be that anomalies in ^{26}Mg might document the previous existence of ^{26}Al? Indeed, meteorite CAIs show ^{25}Mg/^{24}Mg identical to Earth rocks but ^{26}Mg/^{24}Mg higher than terrestrial values. Moreover, in a series of CIAs, ^{26}Mg/^{24}Mg correlates directly with stable ^{27}Al/^{24}Mg—it seems clear that the CAIs formed while ^{26}Al was still active or 'extant'. So, here is further evidence that the earliest Solar System materials were crystallizing within 10 Ma of the formation of the elements.

^{107}Pd (palladium) decays to ^{107}Ag (silver) with a half-life of 9.4 Ma, so roughly an order of magnitude longer than ^{26}Al. Some iron meteorites have ^{107}Ag/^{109}Ag up to eight times higher than the Earth value and again, there is a correlation between the daughter isotope, ^{107}Ag and a stable isotope of the parent element ^{108}Pd. The iron meteorites clearly formed within the first few tens of millions of years of the Solar System.

The next step up in time is to ^{129}Xe (xenon), which is produced by the decay of ^{129}I (iodine), which has a 23-Ma half-life. ^{129}I is one of the longer-lived fission products associated with accidental and deliberate emissions from nuclear weapons and reprocessing plants. Any primordial ^{129}I decayed away within the first 200 Ma of the Solar System, but ^{129}Xe could preserve a record of original ^{129}I. The details are complex but the Earth's mantle (measured in volcanic rocks that originated in the mantle) and atmosphere have different ^{129}Xe/^{130}Xe and carry the implication that the Earth formed within the first 100 Ma of the Solar System.

For me though, the most astonishing observation comes from tungsten (W) isotopes. There is an isotope of the element hafnium (^{182}Hf) that has a half-life of about 13 Ma and which decays to ^{182}W. Tungsten and hafnium have very different geochemical affinities. Hf is lithophile (silicate-rock-loving) but W is both chalcophile (sulfur-loving) and siderophile (iron-loving). The Earth's mantle is mostly magnesium silicate and its core is mostly iron sulfide. At some stage the mantle and core separated from a homogeneous mixture. At that point, tungsten would go into the core but hafnium would go into the mantle. If this separation happened while ^{182}Hf were still active, the majority of the Earth's tungsten would go into the core but the Earth's mantle would then evolve 'new' tungsten from the decay of ^{182}Hf. The core tungsten is a stopped clock because all the hafnium stayed in the mantle. The mantle hafnium keeps ticking to produce new tungsten that is then richer in ^{182}W.

We don't have samples of the Earth's core because volcanoes don't sample the Earth so deeply—although there are suggestions that very deep portions of the mantle are occasionally transported to the surface by volcanic activity. However, we can compare the mantle-derived basalts erupted by volcanoes with the meteorites that we believe represent the starting conditions for Earth. It turns out that ^{182}W/^{183}W is about two parts in 10,000 (i.e. 0.02 per cent) higher in the Earth's mantle than in chondritic meteorites. So, the Earth's mantle and core appear to have separated while ^{182}Hf was still active (i.e. within about 80 Ma of the formation of the Solar System).

So, using isotopes that are the daughters of isotopes that no longer exist, isotope geochemistry has been able to assemble a chronology for the formation and growth of the Solar System and the Earth that is both exquisite and astonishing. Our Solar System began around 4.5 Ga ago, within a few million years the earliest solid minerals crystallized, after about 10 Ma small astronomical bodies had formed, and within 100 Ma the planets had not only formed but had differentiated into bodies that were fundamentally similar to the Earth we live on and the planets we gaze at on a clear night.

Epilogue

Since its inception at a dinner party in Glasgow shortly before World War II, the concept of isotopes has constantly been at the forefront of human endeavour. As we have seen, isotopes have been pivotal in many of the major developments of the past century. From cures for some cancers through to some of the most destructive weapons of mass destruction via understanding the very origins of our own species and Solar System to recognition of human-induced climate change, Soddy's concept of isotopes has had a profound effect in shaping the world in which we live. Frederick Soddy may have worried about the Pandora's box he had opened, and 20th-century history surely vindicates his concern, however, many peaceful and constructive uses of isotopes and their natural abundances have also been developed. I hope this *very short introduction* will leave you motivated to learn more of this fascinating world in which we continue to use isotopes to the benefit of mankind and our environment. Whatever your age or background, isotopes will impact your life and will influence your biggest life decisions. My journey through the sub-atomic world of isotopes has allowed me to understand a little of the exquisite complexity of our natural world. While your encounters with isotopes and isotopic science may be more routine, I hope your trip will be equally rewarding—bon voyage.

Further reading

F. Albarède, *Geochemistry: An Introduction* (Cambridge University Press, 2009).

C. Allègre, *From Stone to Star: A View of Modern Geology* (Harvard University Press, 1994).

S. Bowman, *Radiocarbon Dating* (University of California Press, 1990).

G.B. Dalrymple, *The Age of the Earth* (Stanford University Press, 1994).

G.B. Dalrymple, *Ancient Earth, Ancient Skies: The Age of the Earth and its Cosmic Surroundings* (Stanford University Press, 2004).

A.P. Dickin, *Radiogenic Isotope Geology* (Cambridge University Press, 2005).

T.J. Dunai, *Cosmogenic Nuclides: Principles, Concepts and Applications in Earth Surface Sciences* (Cambridge University Press, 2010).

G. Faure, *Principles of Isotope Geology* (Wiley, 1986).

B. Fry, *Stable Isotope Ecology* (Springer, 2006).

H.E. Gove, *Relic, Icon or Hoax? Carbon Dating the Turin Shroud* (CRC Press, 1996).

C.G. Herbert and R.A.W. Johnstone, *Mass Spectrometry Basics* (CRC Press, 2002).

J. Imbrie, A Theoretical Framework for the Pleistocene Ice Ages. *Journal of the Geological Society.* London **142** 417–32, 1985.

J. Lilley, *Nuclear Physics: Principles and Applications* (Wiley, 2010).

M. Lynas, *Six Degrees: Our Future on a Hotter Planet* (Harper Perennial, 2008).

L. Merricks, *The World Made New: Frederick Soddy, Politics and Environment* (Oxford University Press, 1996).

R. Mitchener and K. Lajtha (eds) *Stable Isotopes in Ecology and Environmental Science* (Blackwell, 2007).

S.R. Taylor, *Solar System Evolution: A New Perspective* (Cambridge University Press, 2008).

W.M. White, *Isotope Geochemistry* (Wiley, 2015).

T. Zoellner, *Uranium: War, Energy, and the Rock that Shaped the World* (Penguin Books, 2010).

Index

R

radioactive decay xv, 2–4, 8–9, 13, 28, 32–4, 52, 82, 86, 90–1, 93, 95, 100

radiocarbon 3, 10–16, 21–2, 35, 63–4

radiostrontium, ^{89}Sr, ^{90}Sr 9, 24, 32, 35, 47, 101

Ramsay, William xiii

rare earth elements 5, 55, 108

Reed, Lou 23

reef-dwelling fish 21

rubidium, Rb 25, 33, 109, 111, 114, 116, 120

RuBisCo 19

Rutherford-Bohr model 1

S

samarium-147, ^{147}Sm 4–5, 111

Scottish Universities Environmental Research Centre (SUERC) 21

SMOW 18

Soddy, Frederick xiii, xv, 1–2, 52, 124

Solar System xv, 5, 24, 81, 90, 94, 108, 118, 120–3

spontaneous fission 3, 8

strontium
^{87}Sr/^{86}Sr 25, 109–11, 114–16, 120–1
strontium isotopes 24

T

technetium 5, 35, 49

Todd, Margaret xiii

tooth enamel 25–6, 78

tritium 2–3, 35

U

University of Glasgow xiii

uranium xv, 5, 9, 12, 39–40, 63, 90, 101–5, 118–19
uranium-234, ^{234}U 40, 63, 98–100
uranium-235, ^{235}U 9, 40, 63, 90, 94, 99, 101–3, 105
uranium-238, ^{238}U 4, 40, 90, 94, 99–100, 102, 118

V

V-PDB 18

V-SMOW 18

W

Walk on the Wild Side 23

Warhol, Andy 23

Index

SOCIAL MEDIA
Very Short Introduction

Join our community
www.oup.com/vsi

- Join us online at the official Very Short Introductions
  Facebook page.
- Access the thoughts and musings of our authors with our
  online **blog**.
- Sign up for our monthly **e-newsletter** to receive information
  on all new titles publishing that month.
- Browse the full range of Very Short Introductions online.
- Read **extracts** from the Introductions for free.
- If you are a teacher or lecturer you can order inspection
  copies quickly and simply via our website.

ONLINE CATALOGUE
A Very Short Introduction

Our online catalogue is designed to make it easy to find your ideal Very Short Introduction. View the entire collection by subject area, watch author videos, read sample chapters, and download reading guides.

http://fds.oup.com/www.oup.co.uk/general/vsi/index.html